access to geography

COASTS *and* COASTAL MANAGEMENT

Michael Hill

Hodder Murray

A MEMBER OF THE HODDER HEADLINE GROUP

Acknowledgements
The publishers would like to thank the following individuals, institutions and companies for permission to reproduce copyright illustrations in this book:

All photos by the author; *Fundamentals of the Physical Environment* by D Briggs, P Smithson and K Atkinson, 1997, Routledge, pages 32 and 45; *Coastal and Estuarine Management* by Peter W French, 1998, Routledge, page 7.

The publishers would also like to thank the following for permission to reproduce material in this book:

The Estate of Anne Ridler for an extract from *Zennor* by Anne Ridler used on page 76.

Note about the Internet links in the book. The user should be aware that URLs or web addresses change regularly. Every effort has been made to ensure the accuracy of the URLs provided in this book on going to press. It is inevitable, however, that some will change. It is sometimes possible to find a relocated web page by just typing in the address of the home page for a website in the URL window of your browser.

Orders: please contact Bookpoint Ltd, 130 Milton Park, Abingdon, Oxon OX14 4SB. Telephone: (44) 01235 827720. Fax: (44) 01235 400454. Lines are open from 9.00 to 5.00, Monday to Saturday, with a 24 hour message answering service. You can also order through our website www.hoddereducation.co.uk.

British Library Cataloguing in Publication Data
A catalogue record for this title is available from the British Library

ISBN: 978 0 340 84638 4

First Published 2004
Impression number 10 9 8 7 6 5
Year   2010 2009 2008 2007

Cover photo: The Rock Islands, Palau (by Michael Hill)
Produced by Gray Publishing, Tunbridge Wells, Kent
Printed in Malta for Hodder Murray, an imprint of Hodder Education, a member of the Hodder Headline Group, an Hachette Livre UK Company, 338 Euston Road, London NW1 3BH

Contents

1 Factors Influencing the Form of Coastlines

O ye! who have your eye-balls vex'd and tired,
Feast them upon the wideness of the sea;
O ye! whose ears are dinn'd with uproar rude,
Or fed too much with cloying melody –
Sit ye near some old cavern's mouth, and brood
Until ye start, as if the sea-nymphs quired.

The Sea, John Keats

1 Introduction

Coastlines and the sea are of tremendous importance to human societies. Over 75% of the Earth's surface is covered by seas and oceans. Furthermore, 50% of the human population lives on coastal plains and in other locations within easy access to the sea. There are numerous human activities that involve the sea, some more specialist such as the fishing industry and the oil industry, others involving a much greater proportion of the population, such as recreation and leisure. Trade has perhaps been the single most important human use of the seas for the last few millennia and it continues to be one of the world's main economic activities, which requires continual modification of the coastline for the construction of ever-larger port facilities. With increasing trade, tourism, population growth in Less Economically Developed Countries (LEDCs) and greater demand for holiday and retirement homes in More Economically Developed Countries (MEDCs), the pressures upon the coast are greater than ever. The image of the untouched coastline is deeply rooted in the human imagination, in art and in literature, which makes the need to conserve as much coastline as possible a great challenge to both present and future generations.

2 Terminology of the Coastline

In English everyday usage there is no real distinction between the terms 'coast' and 'shore'. The dictionary's definition of a coast is:

the side of the land next to the sea, the seashore

and the definition of a shore is:

the land bordering on the sea, a large lake or river.

For geographers, geomorphologists, oceanographers and marine biologists, clearer distinctions need to be made as the coastline represents a whole series of different zones in which specific conditions prevail according to tides, the depth of sea, wave action and other factors. Figure 1 illustrates how these zones relate to one another. As the diagram shows, the term 'coastal zone' applies to the whole area stretching from the furthest point inland where the sea has some geomorphic impact, out to the deeper sea where the effects of the marine processes upon the land are negligible.

In contrast, the word 'shore' is used to identify the subdivisions of the coastal zone, and falls into four categories: **backshore**, **foreshore**, **inshore** and **offshore**. The backshore is the area between the mean high-water mark and the landward limit of marine activity, and is normally associated with changes that take place only during storm periods. The foreshore is the area that lies between the mean high- and low-water marks; this is the most important zone for marine processes in times not influenced by storm activity. The inshore area lies between the low-water mark and the **wave base** (the point at which the water is sufficiently deep for the waves to have little impact upon the land beneath it). In this area wave activity is more limited than in the foreshore. The offshore area is that lying beyond the wave base, in which activity is still more limited and comprises mainly of the deposition of very fine sediments.

Three distinctive zones according to the types of wave action also appear in Figure 1. The **breaker zone** is the furthest out, lying in deeper water where rolling waves start to break. The **surf zone** is that where the breaking waves usually encounter a shallower gradient and have a strong energy impact on the shore, producing a great deal of turbulent water. The **swash zone** lies upon the upper part of the shore where a broken wave pushes upwards onto the shore as swash and then returns gravitationally back down again towards the sea as **backwash**.

3 Waves

Waves are the most fundamental and dynamic force behind the moulding of coastlines. The work they carry out and the force at which they hit the shoreline depend upon the whole character of

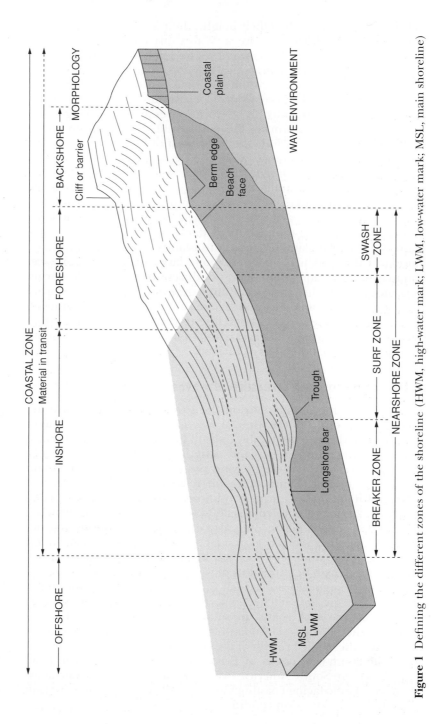

Figure 1 Defining the different zones of the shoreline (HWM, high-water mark; LWM, low-water mark; MSL, main shoreline)

individual waves, including their height, their length, their frequency and the angle at which they break. Waves vary considerably in their power according to weather conditions; under storm conditions wave action can be particularly dramatic, whereas under fair weather they have far less impact upon coastlines. Climate is thus a major influence upon waves and what they can do to coastlines. Parts of the world where coasts are frequently hit by storm waves are those lying in the tracks of depression storms in the mid-latitude belts with prevailing westerly winds. These include the north-west of Europe (Norway, the British Isles, Atlantic France, Galicia in Spain), the north-west of North America from Oregon to Alaska, southern Chile, and the southern coasts of Australia and New Zealand. Regions within the tropics and subtropics that experience tropical storms, such as hurricanes and typhoons may not receive such frequent storm waves but they are often subject to much more damage. Thus, every few years, places such as Bangladesh, Japan, the islands of the Caribbean and Florida in the USA may experience one large surge of storm waves that may have the equivalent energy of dozens of mid-latitude storms. In some parts of the world coastlines experience far fewer storms; these include the more enclosed sea areas, such as the Mediterranean Sea and Red Sea, and places very close to the Equator where, within the doldrums, storm-wind tracks are absent, e.g. the coast of Kenya and the Pacific island nation of Kiribati.

a) The formation and structure of waves

Waves are moving undulations on the surface of the sea that result from the drag effect of the wind. Waves appear as a series of troughs and crests, and vary considerably in both height (the vertical distance between trough and crest) and length (the horizontal distance between two crests). When the wind has been blowing at a constant speed for a period of time, waves increase in size. Off the Atlantic coast of Britain, under stormy conditions, wave heights may be commonly 10–20 m and wavelengths 200–300 m. Wave velocity, which is expressed in metres per second, also depends upon the strength of the wind and other meteorological factors. During stormy periods off the British coast, wind velocities may be as great as 15 m per second.

One of the main determinants of wave size is **fetch**. This is the size of the area of open sea over which the wind may blow without interruption. The greatest fetches experienced in the British Isles are those along the west coasts, in such places as south-west Ireland, Cornwall and north-west Scotland, which are several thousands of kilometres, the whole breadth of the Atlantic Ocean. Some waves that reach the south-west coasts of Britain may, in fact, originate from Cape Horn in South America, which is around 13 000 km away. Britain's Atlantic-facing coastlines are typically very rugged as a result of the frequency of storm conditions.

In deep seas and oceans under normal conditions, water particles do not move great distances; as waves pass over the water surface, the individual water particles move in an oscillatory motion. Particles close to the surface describe larger circles and those further down much smaller ones; when a certain depth is reached particles are no longer in motion but remain static and unaffected by surface wind movements. As a result of this, waves in deep seas have no impact upon the morphology of the sea-bed. These deep sea waves are often known as **swell waves** and any objects floating in them will describe a circular movement and appear to be just bobbing up and down as the wave passes; any lateral movement of floating objects is very slow unless conditions are stormy.

When storms occur at sea, more forceful waves are created by the friction of the wind blowing across the sea surface. These are known as **sea waves** or **forced waves** and, because of the greater amount of energy involved, have a shorter wavelength and a greater wave height than swell waves. When they leave their area of origin and move towards the open oceans they may quickly loose energy and height and become swell waves. If, however, they are generated close to the shore, they may have a dramatic effect in pounding the coastline and bringing about rapid geomorphological changes, such as cliff erosion and the creation of storm beaches.

Swell waves, when they reach shallower water, take on rather different characteristics from those they have in deep water. The sea-bed starts to interfere with their motion and the oscillatory movement becomes more elliptical because of friction between the water and the sea-bed. This process is known as **shoaling**. As a wave comes closer to shore, the wavelength and velocity both decrease but the wave height consequently increases. Once the wave height is greater than the water's depth, the wave becomes unstable and breaks. At this stage the wave is capable of transporting material and is often referred to as a **translatory wave**.

b) The classification of wave types

In addition to the different types of waves mentioned above, waves can also be classified according to their geomorphological action and by the way in which they break upon the shore; both of these are important in the understanding of their impact upon coastlines.

i) Constructive and destructive waves
The basic distinction between constructive and destructive waves is that the former are associated with the processes of deposition and the latter with the processes of erosion. Figure 2 shows the main differences between them. Constructive waves are longer and lower, and break upon a more gently sloping shoreline, such as a beach, with a frequency of around every 10 seconds. The nature of this type of wave

(A)

(B)

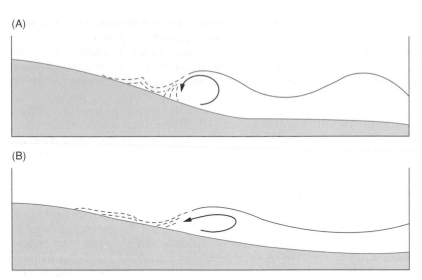

Figure 2 Constructive and destructive waves

is to transport and deposit material up onto the beach, because it has a strong **swash** (which carries sediments up slope following the breaking of the wave) and a weaker **backwash** (which returns some sediments back downslope to the sea as a result of gravity). By contrast, destructive waves have a shorter wavelength and a greater height, and hit a typically steeper shoreline with a frequency of around every 5 seconds. The plunging nature of these waves has an erosive effect and, although it has only a weak swash, it has a strong backwash that carries eroded materials back into the sea.

ii) The different types of breaking waves

As breakers reach the shore there are four main ways in which they may break and this will affect whether they are more likely to cause erosion or deposition. These four types of waves can be identified by their shape, and they are:

- **spilling waves**, which break onto gently sloping shorelines and are characterised by a great deal of foam and turbulence as their crests spill and cascade downwards. They tend to develop over wide, open beaches and their energy is quickly dissipated and any erosion and deposition tends to be held in a state of equilibrium
- **plunging waves**, which are steep-fronted waves that break onto relatively steep shores. As they break and curl over with great energy they have strong erosive power and a strong backwash
- **collapsing waves**, which start to break at a fairly steep angle but their crests then collapse, followed by the rest of the wave. When

they break much of their energy is reflected back into the sea; they therefore are associated with strong backwash properties
- **surging waves** are very similar to collapsing waves and with similar properties; they differ in collapsing before the crest of the wave reaches such a steep angle.

c) Wave refraction

When incoming waves approach the shore they may be **refracted**, i.e. made to change their direction. This may occur as a result of various circumstances, including:

- waves coming in at an angle that is oblique to the shore
- the coastal topography that the waves are approaching is varied
- the sea-bed topography below the waves is varied.

A typical example of wave refraction is shown in Figure 3, where waves are approaching a sequence of headlands and bays. The dashed lines represent the wavefront and show the refraction taking place as a result of the local topography. The arrows represent hypothetical lines called **orthogonals**, which run at right angles to the wavefronts

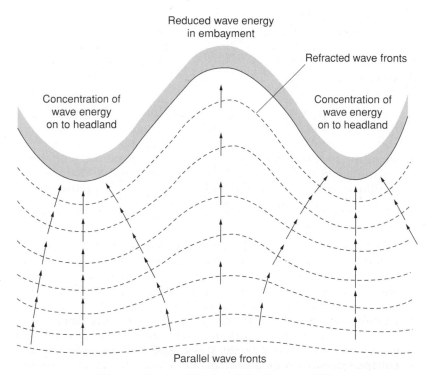

Figure 3 Wave refraction (after French 1997)

and show the directions that individual waves are likely to follow. These orthogonals converge towards certain points and it is here that the wave energy becomes most concentrated. From the diagram it is clear that headlands are more vulnerable to wave attack and, by contrast, bays remain relatively sheltered from wave energy.

d) Tsunami or 'tidal waves'

The word '**tsunami**' is the Japanese for 'harbour wave'; it is therefore rather misleading but no doubt has its origins in the fact that harbours would have been destroyed when tsunami hit them. The English term 'tidal wave' is equally misleading, as tsunami have nothing to do with the changing patterns of the tides. Tsunami are very large waves up to 30 m in height and therefore generally well above the size of normal storm waves. They are activated by earthquakes, other movements in the lithosphere, as well as other factors that cause the seas to swell way beyond their normal wave patterns.

Although the most frequent cause of large tsunami has historically been earthquakes, volcanic eruptions and landslides are also responsible for their development. They mostly occur within the basin of the Pacific Ocean where they are activated by earth movements within any of the subduction zones that surround the ocean or within any of its various island arcs. An earthquake on one side of the Pacific may generate tsunami that take as long as 8 hours to reach the other side. The Hawaiian Islands are particularly vulnerable as they lie in the mid-Pacific and can receive earthquake-generated tsunami from both the eastern parts of Asia and the west coast of the Americas.

One of the greatest recorded tsunami followed the massive Plinian eruption of the volcano Krakatoa off the west coast of Java (Indonesia) in 1883. This wave reached a height of 35 m and inundated many islands of the East Indies; 2800 people were washed out to sea in the port of Batavia (now Jakarta) alone and the total death toll is estimated to have been between 36 000 and 80 000. Several small islands such as Midah and Siuku disappeared altogether under the wave and all their inhabitants were drowned.

Japan's biggest bronze statue of the Buddha, which sits 11.36 m high in the town of Kamakura near Tokyo, is in the open air rather than inside a great hall as are similar statues elsewhere in the country. The Buddha is over 3 km inland from the Sagami-Wan bay and lost the wooden temple-hall that housed it as the result of an earthquake-generated tsunami that hit the town in 1495. The bronze statue was too heavy to be carried away by the tsunami.

Landslides often generate tsunami. There is evidence that an enormous underwater volcanic rock-slide may have taken place in the Hawaiian Islands about 105 000 years ago; the resulting tsunami would have been even greater than the Krakatoan one. On a more modest scale, a piece of lava of equivalent size to a three-storey building fell into the sea from the erupting volcano Stromboli in Italy in December 2002. The wave inundated all of the houses along the seafront of the island's main settlement.

Ice falls can also generate tsunami; disturbances have been witnessed around the coasts of both Greenland and Antarctica as a result of iceberg calving.

In remoter geological times, millions of years ago, there is a high probability that asteroid impacts could have generated gigantic tsunami far greater than any in recorded history. In recent years there has been a lot of attention paid to what impact an asteroid hitting the earth might have in the future and it has been generally assumed that it will fall upon the land. Given the amount of the planet's surface covered by sea, it would be more likely to create a tsunami that would drown vast areas of land rather than a terrestrial impact crater.

4 Tides

Along with waves, tides are the other major type of phenomenon that controls the interactions between sea and shore. In certain parts of the world both these aspects of the natural environment are of equal importance. In others, such as those places that experience very high tidal ranges, tides may be of more significance in determining the nature of coastlines than waves. Also, certain natural features are more influenced by tides than waves and these include many low-lying environments such as estuaries, salt marshes and mudflats.

a) The causes of tides

Tides are effectively waves with very long wavelengths – which stretch half way round the globe. They are caused by the effects of the gravitational pull of the sun and the moon (and also to a very marginal extent by that of some of the nearer planets) upon the earth's oceans and other water surfaces. These effects cause the oceans to have a **tidal bulge** where water is deeper and a **tidal trough** where it is shallower. As the sun and moon pass overhead the tidal bulge travels with them and this explains why the tidal cycle is diurnal with two high tides and two low tides a day in most coastal locations. The cycle does not coincide exactly with our planetary day, as there is a lapse of 6 hours and 13 minutes between high and low tides; this explains why high and low water times on tide tables get slightly later each day.

The magnitude of tides depends upon the relative positions of the sun to the earth and moon within their orbital cycles. When the sun and moon are aligned and exert their gravitational pulls in the same direction the tides are highest and known as **spring tides**. When the sun and moon are at 45° angles in relation to the earth the tides are at their lowest and known as **neap tides**. This tidal cycle coincides approximately with the lunar month with roughly 14 days between two spring tides or two neap tides.

b) Amphidromic systems

The pattern of tides within sea areas is made more complicated by the **Coriolis effect**, which is caused by the rotation of the earth. Any system moving across the earth's surface, such as tides or winds, is deflected by the Coriolis effect. This is anticlockwise in the Northern Hemisphere and clockwise in the Southern Hemisphere. Thus, tidal movements travel in a circular pattern and this is known as **amphidromic motion**. At the centre of an amphidromic system, rather like the hub at the centre of a wheel, is a point at which the tidal range is zero: the **amphidromic point**. Outwards from this point the tidal ranges become more extreme. The lines radiating outwards from the amphidromic points are known as **cotidal lines** and they represent the differences in time of high tide in hours.

Exactly where amphidromic points are located depends upon the configuration of land and sea, the ocean depths and the sea-floor topography.

c) Tidal ranges

Davies in his work 'A morphogenic approach to the world's shore-lines' defined three types of coastal environments based upon their tidal ranges.

- **Macrotidal** areas are those that experience a tidal range of more than 4 m. This tends to occur where the continental shelf is wide and the configuration of the coast helps to amplify the wave height. Many parts of north-west Europe and the north-eastern North America fall into this category, and include the places with the two highest tidal ranges in the world – the Bay of Fundy in Canada (with a range up to 15 m) and the Bristol Channel (with a range up to 13 m). Other coastlines of the world that are macro-tidal include that of southern Argentina, north-west Australia and Bangladesh. In these macrotidal environments tidal change plays a major role in coastal processes.
- **Mesotidal** areas have a tidal range of between 2 and 4 m. Coastlines that come into this category include that of the countries bordering the South China Sea (e.g. Malaysia, Indonesia, Vietnam), East Africa and the Caribbean. Mesotidal coasts often have equal geo-morphological roles being played by tides and wave action.
- **Microtidal** areas have a tidal range of less than 2 m. They include the east coast of Australia and the west coast of Africa. Along these coastlines wave action is much more important than the role of tides in the development of coastal features.

CRAVEN COLLEGE

The large tidal ranges experienced in some of the oceans are not to

Tides

The large tidal ranges experienced in some of the oceans are not to be found in the smaller enclosed seas of the world. The Mediterranean, with a surface area of $2\,505\,000\,\text{km}^2$ and a narrow outlet into the Atlantic Ocean through the Straits of Gibraltar, is effectively the largest inland body of water in the world. Its tidal ranges are very modest in comparison with the major oceans because the gravitational pull of the sun and moon have a much smaller body of water on which to operate, therefore the Mediterranean marine environment is a microtidal one. Average tidal ranges within the Mediterranean are around half a metre, but they vary considerably from place to place. The funnelling effect seen in the Severn Estuary and the Bay of Fundy is also found in the Mediterranean. Within Italy, for example, the highest tidal ranges are experienced at Trieste, which is located at the landward head of the Adriatic Sea, the Mediterranean's narrowest branch. Trieste's average spring tides have a range of around 95 cm, in sharp contrast with ports such as Catania and Taranto, which are located in much more open areas of the Mediterranean and have ranges of just over 20 cm during spring tides.

d) Tidal surges

Surges are changes to the normal predicted astronomical tidal patterns that occur during abnormal meteorological conditions. Two major factors contribute to the formation of surges:

- strong winds, which can change the height of coastal waters by 2–3 m
- changes in atmospheric pressure patterns, such as intense depressions, that alter the height of coastal waters by around 1 cm per millibar.

Surges can only really occur in relatively shallow waters but can have a profound effect on the coastlines that they reach. There are two main types of surges: **positive surges**, which lead to abnormally high tides and may cause flooding, and **negative surges**, which lead to very low tidal conditions and may be a hazard to navigation, causing ships to run aground.

The lands surrounding the North Sea suffered from a great storm surge in January 1953, which coincided with a spring tide that was around 3 m higher than normal. The worst-affected areas were southeast England and the Delta region of south-west Holland, where 1800 people lost their lives. As a result of the damage caused by the surge, both Britain and the Netherlands embarked upon new flood defence schemes. In Britain the Thames Barrier was created in order to prevent the large amount of low-lying land in London from being flooded from the sea; the Barrier eventually came into operation in 1983. In the Netherlands the much more complex Delta Scheme was created. This linked the various islands in the mouth of the River

Scheldt with a series of dams to prevent surges moving up the distributaries of the river and flooding cities and densely populated rural areas.

e) Tidal bores

Tidal bores are large waves that surge up rivers and estuaries with the incoming tide, effectively forcing a reversal of the river's flow. They only occur where certain natural conditions are right. The configuration of the estuary is important, but above all else it must be **hyper-synchronous**. This means that the convergence of the sides of the estuary upstream should be rapid enough for the incoming tide to outweigh any of the effects of friction that might be created by the estuary's bed.

There are some 60 rivers in the world that experience regular tidal bores that coincide with the spring tides. They are by no means restricted to regions that experience high tidal ranges, as they include the Batang Lupar in Sarawak, Malaysia and the Qiantang River in China, both of which flow into mesotidal waters. The Qiantang has the biggest tidal bore in the world, which regularly reaches 8 m in height and there has been a long history of 'bore watching' on the river going back over 2000 years. In 1993 when the bore reached over 9 m, there was extensive flooding and over 100 people lost their lives.

Tidal bores are more common in macrotidal areas of the world and include those of the Araguari River, a tributary of the Amazon in Brazil, and the Bay of Funday, Nova Scotia in Canada. The funnel-shaped Bay of Funday only has a small tidal bore, although it has the highest tidal range in the world. However the water becomes more concentrated and channelled along its tributaries, which have significant bores; these include the Stewiacke, Salmon and Shubenacadie Rivers. In Nova Scotia not only does 'bore watching' take place, but there are also various adventure sports that take advantage of the waves.

In Britain the highest bore is found on the River Severn, which has the second biggest tidal range in the world and is funnel-shaped like the Bay of Fundy. The bore reaches on average 2–3 m in height during the spring tides and is a favourite with local surfboarders. Other rivers that have bores in Britain include the Parrett in Somerset, which drains into the lower part of the Severn where it has already become the Bristol Channel.

5 Geology, Lithology and Relief

The physical nature of the land has as much impact upon coastal forms as the various factors that influence the work of the sea. At a regional or national level, plate tectonics and geology influence the overall trends and form of the coastline. Geological trends create two distinctive types of coasts.

- **Atlantic** or **discordant coastlines**. These are coastlines where the general geological trends of mountains and hills are at right angles to the shore. Thus, they are typified by headland and bay topography, either on a localised scale or on a much larger, national or regional scale. At the large scale, most of southern Greece, and in particular the three peninsulas that form the coastline of the Peloppenes, and most of the west coast of Ireland are typical Atlantic coastlines. At a smaller scale, the coastline of north-west Devon within Barnstaple Bay and the central part of the northern· coast of São Miguel island in the Azores are dominated by discordant headland and bay topography.
- **Pacific** or **concordant coastlines (or 'Dalmatian' coastline)**. These coasts have a geological trend that is parallel to the shoreline. They may be dominated either by relatively straight, rocky coastlines or have a series of long, narrow islands parallel to the shore. Most of the eastern shore of the Adriatic branch of the Mediterranean Sea has this type of coastline. Stretching from the Istrian Peninsula in Croatia, through Montenegro and Albania to the island of Corfu in Greece, the Dalmatian coast is probably the best example of a concordant coastline in the world, and indeed its name is sometimes used to describe this type of coast. At a much smaller scale, the stretch of Dorset coast from Osmington Mills to Kimmeridge Bay is concordant.

The 'Atlantic type' coastline on the north coast of São Miguel, the Azores

Another way in which geology has an important effect upon the nature of the coastline is in the rates of erosion of different types of rocks. Measurements of cliff recession throughout the world have enabled rates of erosion for different rocks to be quantified. Average rates of recession for selected rocks are as follows:

- granite: 1 mm/year
- limestone: 1 mm–1 cm/year
- shale: 1 cm/year
- chalk: 1 cm–1 m/year
- glacial drift deposits: 1–10 m/year
- volcanic ash: > 10 m/year.

The ways in which rock types and geological stratification influence the coastline on a small scale and in particular determine cliff profiles will be looked at in detail in the next chapter.

6 Weathering and Mass Movement

Although the wind, waves and tides all play a great role in modifying the rocks, along the shingle and sand coastlines of the world, the roles of weathering and mass movement may be even more important in moulding the shoreline. Weathering or subaerial erosion is important along rocky coastlines in determining the overall shapes and profiles

Hopewell rocks exposed at low tide in the Bay of Fundy, Canada, which has a 15 m tidal range

of cliffs, and this will be looked at in more detail in the next chapter. Weathering involves the interaction between certain physical, chemical and biological phenomena with the soil or rock surfaces of the lithosphere. The two biggest controls upon which forms of weathering take place are therefore climate and geology. Each climatic belt of the world will experience combinations of weathering processes taking place along their coastlines. For example, in cold temperate and polar climate zones, **freeze–thaw** action is likely to be highly active along rocky coastlines, whereas cliffs along desert coastlines will be more subject to waterless forms of thermal fraction. In humid tropical locations, the abundance of vegetation and fauna lead to much more active biological forms of weathering, whereas in humid temperate locations such as the British Isles it is often the sheer intensity of rainstorms that is the main cause of the weathering and consequent failure of coastal slopes.

Along limestone coasts, **chemical weathering** may be one of the most important processes at work; direct contact with sea water or sea spray charged with carbonic acid and other caustic solutions lead to the gradual wearing away or the pock-marking of the rocks. Chemical weathering is also significant along some arid coastlines, particularly along hyperarid coasts such as that of the Atacama where there are great deposits of mineral salts. **Biological weathering** by plant roots is particularly active on vegetated upper slopes of cliffs. Burrowing animals cause the biological weathering of soft rocks such as clay and also disturb coastal sand dunes. In tropical limestone habitats crabs and other sea creatures are responsible for a lot of the wearing away of features such as shore platforms over which they scavenge both at high tide and low. If we add the human population to the biological weathering processes, a wide range of both accidental and deliberate actions can be brought into the equation from trampling over sand dunes, through miscalculated coastal protection schemes to replacing natural coastlines with built environments.

The processes of **mass movement** are intricately bound up with geological structure of the coastline, the weathering processes that are taking place upon it, and the controls imposed by climate, past and present. In high-latitude periglacial regions, rockfalls from the material loosened by freeze–thaw action and solifluction lobes frequently form part of the coastal environment – although only visible in the warmer months of the year. In cool and warm temperate regions such as Britain and the countries of the Mediterranean, a wide range of mass movement type can be found around the shores: soil creep is commonly found on the gentle grassy upper slopes above cliff faces, much more dramatic siding and slumping is found where rocks are soft but interbedded, and rock falls are common on the more sheer and vertical stretches of cliffs. Along humid tropical coastlines dense vegetation cover protects the coastline from potential mass movements, but when this is removed either by natural or human actions, heavy rainfall will have a dramatic effect in triggering off landslides, slumping and rockfalls.

7 Human Activities Along Coastlines

No other creatures have had as much impact upon coastlines as human beings. The long history of human colonisation of coastlines and the wide range of human activities involving the sea and the coastline have had an immense influence upon coastal locations. Human activity is highly localised and vast stretches of relatively untouched shorelines still survive in the more hostile environments of the Arctic and Antarctic, the humid tropics and the desert areas of the world.

Increasingly, however, in a globalising world, some of these more remote coastal environments are coming under threat. Population pressures, the quest for minerals and other natural resources, and marine pollution are just three ways in which coastlines are becoming increasingly threatened by human activity.

The different forms of human interaction with the coast could be made into a very long catalogue of activities, but some of the more important ones include:

- the building of harbours, ports and coastal cities
- the development of coastal resorts and leisure facilities
- the development of the fishing industry and associated infrastructure
- the reclamation of coastal marshes to extend the land for farming, industry and other activities
- the development of the oil industry, its ports, tankers and terminals
- the pollution of the seas from industry, farming and oil spillages
- the management of the coastline to stabilise cliffs, slow down erosion and protect property
- the construction of coastal flood prevention schemes
- the use of tides as a means of generating electricity
- the action to conserve important coastlines of great natural beauty and coastal ecosystems of great scientific value.

Many of these themes will be dealt with later on in the book, especially in Chapters 6 and 7.

Summary Diagram

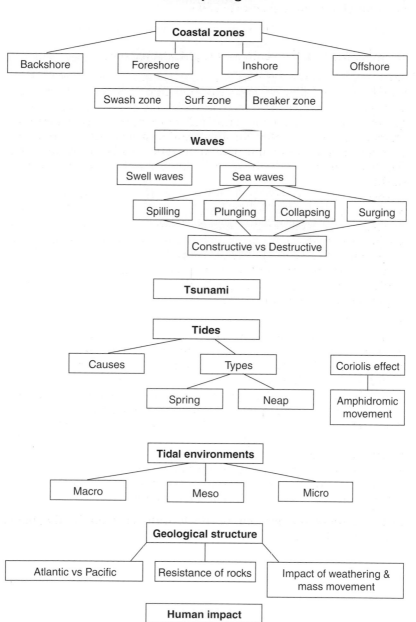

Questions

1. a) Outline the main factors that are responsible for the development of different types of coastlines.
    b) For any two of these factors expain in detail how they can create different coastal landforms.
2. a) What are the main reasons why (i) waves and (ii) tides occur within the oceans?
    b) Assess the relative importance of waves and tides in shaping coastlines.
3. a) Under what conditions do macro- and microtidal environments occur?
    b) What are likely to be the main differences in their coastal characteristics?
4. a) Explain the differences between constructive and destructive waves.
    b) Examine the effects these two wave types have upon the shoreline.
5. a) To what extent extent do geology and mass movement influence coastal landforms?
    b) What impact do humans have upon the coastal environment?

2 Rocky Coastlines and the Processes of Erosion

KEY WORDS

retreating coastline a coastline that is subject to erosion rather than deposition
hydraulic action the erosive force of the impact of waves upon the shore
corrasion erosion caused by waves 'armed' with rock fragments
attrition the wearing down of rock fragments into pebbles and smaller particles
solution the chemical action of sea water on limestone and other soluble rocks
notch the initial groove eroded at the bottom of a cliff face by marine action
cliff profile the cross section of a cliff
shore platform the gently sloping relic feature left behind following coastal retreat which is exposed at low tide; also known as a **wave-cut platform**

Wild, wild the storm, and the sea high running,
Steady the roar of the gale, with incessant undertone muttering,
Shouts the demoniac laughter fitfully piercing and pealing.
Waves, air, midnight their savage trinity lashing ...

Patroling Barnadat, Walt Whitman

1 The Main Processes That Cause Coastal Retreat

Cliffs and the other features of rocky shores are attacked and eroded in a number of different, but inter-linked, ways. There are generally recognised to be four main processes by which the sea erodes coastlines: **hydraulic action**, **corrasion**, **attrition** and **solution**. Some sources also add the term **quarrying** or **cavitation**, but as will be explained below, this is a process that results from some of the other modes of erosion and should not be singled out as being a separate process. Weathering and mass movement also have very important roles to play in determining the shape of rocky shorelines. Weathering is much more active than erosion on the upper portions of cliffs where the effects of the waves rarely reach and therefore is often of equal importance in the development of cliff profiles. Mass movements such as slumping and rockfalls result from both weathering and marine erosion, and consequently also have a profound effect upon the detailed morphology of the coastline.

a) Processes of erosion

Rates of erosion depend upon many factors, most of which were considered in the last chapter. The nature of the waves, their strength, their frequency, their height and the frequency of occurrence of storm conditions are all important elements in the erosive marine environment. Equally important in this interface between the sea and the land is the geological nature of the coastlines themselves: the type of rocks from which they are made, their degree of resistance, their stratification and their stability. The four types of erosion operate within this framework formed by the interrelationships between sea and land; their degree of effectiveness varies considerably, both spatially and temporally.

i) Hydraulic action

This is arguably the most important and most destructive of the four processes of marine erosion and is carried out by the impact of waves breaking upon the rocks of the coastline. It is estimated that normal winter Atlantic breakers hitting the shores of western Britain can exert pressures of 10 000 kg per m² and that during heavy storms the pressures can be as great as 30 000 kg per m². Another idea of the destructive nature of hydraulic action can be gained from observations made following wave impact. Large bodies of sea spray moving into the air can achieve velocities of up to 120 km/hour, whereas more mobile, smaller bodies have been observed to be travelling at an astonishing rate of 280 km/hour.

The effectiveness of the erosive capacity of a wave depends upon where it breaks in relation to the shore, as the water at the crest of a breaker actually has twice the velocity of other parts of the wave. When the breakers hit the shore, it is not as much the weight of the water itself that causes rocks within a cliff to fragment, as the pressure exerted upon pockets of air contained within cracks, joints and other irregularities on the rock surfaces. When air contained within such irregularities is compressed for a few seconds, it is then followed by **pressure release** that takes the form of an explosion and tears the rocks apart. Observations have been made that show that erosion of this sort is most effective where the air pockets are relatively thin.

ii) Corrasion

Also known as **abrasion**, this is in many coastal locations the second most important form of erosion. Rock fragments are picked by the backwash of destructive waves and then thrown against the rock face of cliffs and other features by subsequent breaking waves. The effectiveness of corrasion depends upon the strength of the wave, the angle at which it breaks, the volume, size and angularity of the rock fragments that are thrown against the shore, and the degree of resistance to erosion of the parent material of the shore. Corrasion, like hydraulic action is most effective at the base of a cliff and has negligible impact higher up. The combined work of these two processes can therefore lead to the development of undermined cliffs with overhanging upper faces. Once this **quarrying** or **cavitation** has reached a certain point, the upper part of the cliff becomes unstable and large rockfalls or landslides take place. Quarrying cannot be regarded as a separate form of erosion as it results from the two processes of hydraulic action and corrasion.

iii) Attrition
Rock fragments that have become detached by the quarrying action of hydraulic action and corrasion are worn down into both smaller and more rounded pieces. Currents and tidal movements cause the fragments to be swirled around and to grind against each other. Pebble and cobble beaches are the product of this form of erosion. The process of attrition is most effective in the breaker zone and in places with a high tidal range; there are cycles of which part of the beach is being affected by attrition at a given time of day.

Most erosion is carried out during stormy rather than fair-weather conditions. Hydraulic action and corrasion are in particular made more forceful by storm waves, the former gaining more weight from the sheer volume of water, the latter being given greater ability to pick up a heavier and larger load. Attrition is also more effective in breaking down eroded material when seas are more turbulent – for example under stormy conditions, or where currents are strong.

iv) Solution
Solution is the one erosion process that is not more rapid with the onset of stormy conditions. It is more influenced by sea temperatures, because chemical processes tend to be more effective in warm water than in cold. On the world scale, therefore, solution is more effective in tropical and subtropical waters than in temperate or polar waters. Solution is the chemical attack on limestone by the carbonic acid, which is formed as sea water absorbs carbon dioxide and by the other chemicals dissolved in sea water, in particular the **halites** such as chloride, iodide and bromide ions. Some sources suggest that the solubility of limestone by carbonic acid decreases with higher temperatures. This may be so, but the other chemicals in sea water are more effective in wearing the rocks away, and these do become more active at higher temperatures. Organic acids produced by algae are also present in these waters and may contribute to the solution of limestone.

In the tropics and subtropics there are many coastal landforms that are either unusual or absent in temperate regions, such as the **notch and visor** formations commonly found upon shore platforms in coralline limestone. These are undercut rock pedestals or mushroom blocks, which may reach up to as much as 10 m in height. These features are widely dispersed throughout the world, for example around

Limestone pillars left behind by chemical action, Nauru Island, central Pacific

Mombasa in Kenya, on the Philippine island of Samar and the east coast of Barbados. Along these coastlines solution also creates irregular, jagged pinnacles called **lapiés**, which may be located between the high- and low-water marks and on low-cliff tops, but in the latter case they are as much the product of rainfall action as from sea spray. In certain tropical locations, such as around the coastline of the Pacific island republic of Nauru, vast numbers of pinnacles form landscapes known as **spitzenkarren**.

Also in tropical and subtropical regions, the constant splashing of limestone rocks by the sea spray can lead to deeply eroded and irregular honeycombed and pitted surfaces, for example on many Mediterranean island such as Giannutri and Lévanzo, both in Italy. Small-scale grooves may also be eroded into the limestone surfaces by sea spray being drained off the shore back into the sea. These are common along Mediterranean limestone coasts, for example the coastline around Návplion in Greece, and are known as **rillenkarren**.

In temperate regions, solution rates are higher in the summer and autumn months when sea temperatures are at their highest. Although the pitting of limestone surfaces by sea spray is common, the more developed karst forms of the tropics are much rarer.

b) The role of weathering and mass movement

Weathering or **subaerial erosion** is a much more important process of denudation on the upper parts of cliffs than marine erosion. Rainwater action, the freezing and thawing of moisture trapped within joints and cracks in the rocks, and thermal expansion and contraction of rocks all play a significant role in weakening the upper surfaces of sea cliffs. Which of these processes is most effective in any given location depends both upon climate and the lithology of the cliffs. Freeze–thaw weathering and the mass movement by solifluction are likely to be important processes along coastlines in high-latitude locations. In the temperate regions, raindrop action, some freeze–thaw in the winter and some biological weathering are likely to be common along coastlines, as are a whole range of slides, slumps, falls and creeps. In humid tropical locations, heavy rainfall is the dominant weathering process, as long as the coastline is not protected by vegetation cover – in which case biological action is likely to be more important. Once the vegetation is removed a whole series of different types of rockfalls, landslides and mudflows could occur.

The way in which lithology influences the processes of denudation is examined below in the sections on cliff profiles and cliff failure.

2 Cliffs and their Profiles

Cliffs are the most fundamental and ubiquitous features of rocky coastlines. They vary tremendously in their height, shape and profile. In the British Isles cliff heights range from just a few centimetres in areas dominated by salt marshes, to well over 300 m. The highest cliffs in the British Isles are in the Irish Republic where they reach over 400 m in Dingle Bay; in Great Britain the highest cliffs are at Countisbury near the Devon–Somerset border where the upland massif of Exmoor drops off steeply from 300 m down into the Bristol Channel.

Not all cliffs are actively experiencing retreat under average conditions as they may be protected by other features such as sand banks and barrier islands. Fossil cliffs such as those associated with raised beaches (left behind after a rise in land levels) are also very unlikely to be eroded by the sea.

Cliffs vary in angle from 90° sheer rock faces down to as little as 13°. Often the slopes are not constant but may be made up of a number of different faces, this can be the result of there being gentler debris slopes at the foot of cliff, of the different processes at work on different parts of the cliff face (subaerial erosion on the upper parts and marine erosion on the lower parts), or because of slumping and other forms of slope failure.

a) Classifying cliffs by climate zone

Many attempts have been made to classify cliffs and one of the most useful of these is that made by Davis (1980), which puts them into a broad world climatic context. Davis recognised four main categories of climatic conditions under which cliffs may form:

- **Tropical cliffs**. These retreat at low angles because they are protected from the sea's breaking waves by coral reefs, and the roots of dense vegetation commonly present bind the cliff material together. These tropical bluffs therefore often have dense rainforest coming right down to the level of the sea.
- **Arid cliffs**. In desert regions cliffs are likely to be of a steep angle because there would be little vegetation cover and a lack of coastal deposits brought down by rivers, both of which protect the cliffs from the full force of wave attack.
- **Temperate cliffs**. In mid-latitude regions cliffs experience high-energy waves, particularly in locations where they get the full force of westerly winds and depressions. This means that the effects of marine erosion tend to be dramatic and give rise to steep-angled cliffs.
- **High-latitude cliffs**. Cliffs in polar regions tend to have low slope angles. In the colder months they are protected from marine erosion by the presence of snow and ice, and in the summer months are protected by the large accumulations of debris from periglacial processes.

b) Cliff profiles

Cliff profiles are extremely varied and, like other types of slopes, may be either simple or complex. The forms of cliff profiles reflect the many different factors that influence shorelines, these include:

- the lithology and integrity of the rocks from which the cliffs are made
- whether or not the cliffs are interbedded with layers of rocks of different degrees of resistance to erosion
- the direction in which the strata are dipping, in the case of sedimentary rocks
- the frequency of wave action under stormy conditions
- the role of subaerial erosion on the upper parts of the cliff face.

Figure 4 shows a selection of 15 different types of cliff profiles. Low and, sometimes, rounded profiles, such as that shown as (A), are typical of soft shore materials that have little resistance to erosion and are incapable of developing into a steep slope. Clays, mudstones and glacial materials such as boulder clay tend to develop into these low profiles, which rarely achieve more than about 50 m in height. Boulder clay is particularly easily eroded because of its lack of coherence. Examples of cliffs like these are found along the north coast of Norfolk associated with the Cromer Ridge and along the west coast of the Jutland Peninsula in Denmark where there are extensive moraine deposits.

By contrast high and vertical cliffs, shown in (B), form where there are more resistant rocks such as granite, limestone and hard sandstones; not only may these cliffs achieve heights of over 500 m, but

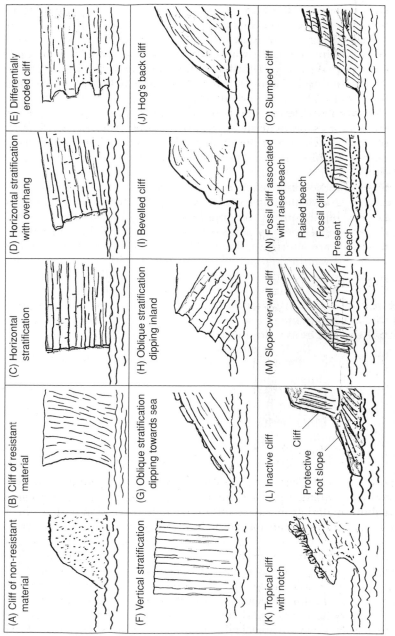

Figure 4 Cliff profiles (note: diagrams are at different scales)

(A) Cliff of non-resistant material

(B) Cliff of resistant material

(C) Horizontal stratification

(D) Horizontal stratification with overhang

(E) Differentially eroded cliff

(F) Vertical stratification

(G) Oblique stratification dipping towards sea

(H) Oblique stratification dipping inland

(I) Bevelled cliff

(J) Hog's back cliff

(K) Tropical cliff with notch

(L) Inactive cliff

Cliff

Protective foot slope

(M) Slope-over-wall cliff

(N) Fossil cliff associated with raised beach

Raised beach

Fossil cliff

Present beach

(O) Slumped cliff

they may also be able to support overhanging upper edges. Examples of these sheer vertical cliffs in Britain include the granite cliffs in many parts of the Penwith Peninsula in Cornwall, the white chalk cliffs along parts of the Sussex and Kent coasts such as those that form the Seven Sisters, near Seaford and the Great Whin Sill (made of highly resistant dolerite), where it reaches the sea and provides a highly defensive site for Dunstanburgh Castle in Northumberland.

Within sedimentary rocks, the nature of the bedding and patterns of folding are important in determining the forms of cliff profiles. To a lesser extent, the direction in which the cracks and joints lie in other types of rocks, such as granite are also of significant in their cliff profiles. Horizontal bedding (C), as in the chalk cliffs of the Seven Sisters in Sussex, enables the formation of high vertical cliffs that lose blocks from their upper portions through weathering where overhangs occur. If the bedding plane is horizontal but dips slightly seawards (D), there are ideal conditions for vertical cliffs with overhangs; these cliffs are highly prone to toppling, as is the situation in the Carboniferous Limestone cliffs at Goven's Head in Pembrokeshire. When there are alternating bands of hard and soft rock, differential erosion and weathering take place, giving the cliff a stepped profile (E), this is particularly noticeable in horizontally bedded rocks. Perfectly vertical bedding planes (F) lead to the formation of high vertical cliffs, but this situation rarely occurs in reality. Where the dip of the bedding plane of the rocks is towards the sea and greater than 20°, there is a gently sloping cliff, but blocks that are loosened by weathering are likely to slide down slope, giving the cliffs a slightly stepped profile (G). Where bedding planes are dipping away from the sea (H), cliffs with a slope from 70° to vertical are likely to form; differential erosion and weathering tend to give such cliffs stepped profiles. Most of these bedding patterns are found in the cliffs at Kilve in Somerset, which are dealt with as a case study later.

A large range of cliff types do not have profiles with constant slopes but, as with other types of slopes such as hillsides, may be made up of more than one slope unit. **Bevelled cliffs** (I) are common throughout the world and simply reflect the fact that weathering is the main group of processes at work on the upper slopes and give them a rounded profile, whereas the dramatic processes of marine erosion produce the much steeper lower slopes. Vegetation cover helps to make the upper slope relatively stable. Where bevelled cliffs form in softer rocks, such as clays and soft sandstones, the overall profile is much smoother and rounded; this is often known as a **hog's back cliff** (J). In tropical locations where the shore is formed of soluble rocks, such as coralline limestone, solution leads to the formation of deep notches at the foot of cliff faces and in some cases these notches become caves. Such tropical cliffs (K) have distinctive breaks in slopes and are well developed in many locations such as a Liliputa Beach on Samar Island in the Philippines and on Efate Island in Vanuatu.

Heavy deposition at the foot of a cliff may create an apron of gently sloping material, which actually protects the older, steeper cliff face from marine erosion, although weathering may continue, particularly on the upper slopes. Cliffs protected in this way (L) are often known as **inactive** or **dead cliffs**, distinguishing them from cliffs that are subject to marine erosion. Deposition from slope failure and solifluction from the Pleistocene ice ages are two ways in which the protective lower slopes have formed; these gently sloping areas are known as **undercliffs**. The cliffs at the head of Rhossili Bay on the Gower Peninsula in South Wales are a good example of steep cliffs protected by a solifluction terrace. Pleistocene deposits may, in some cases, totally overlie old cliff faces and totally bury them, giving the modern coastline a different profile. This type of cliff, known as a **slope-over-wall cliff** (M), results from solifluction materials covering old cliffs and now being undercut by marine erosion; Dodman Point in Cornwall is one of many cliffs areas in Britain formed in this way. Where **raised beaches** occur as a result of tectonic uplift following the Pleistocene ice ages (which will be dealt with in more detail in Chapter 5), there are inactive cliffs forming part of the shore profile (N); such profiles are quite common in some parts of Britain, such as near Start Point in South Devon. Where slumping and slipping forms of mass movement occur, cliff profiles take on a stepped form (O). There may be just one step or a whole series of them, according to the degree of slope stability. Many parts of southern England have stepped cliffs as a result of rotational slumping, e.g. Folkestone Warren in Kent and the coastline near Ventnor, Isle of Wight.

Davies (1980) put sea cliff slope profiles into three categories according to the relative importance of weathering and erosion. Cliffs where marine erosion is much more important than weathering he saw as typically very steep or vertical and retreating as a result of undercutting being followed by cliff collapse. Where both weathering and marine erosion are of similar importance a bevelled cliff profile is likely to emerge. Where weathering and mass movement are more important than marine erosion, the slumped and stepped profile is most likely to evolve.

CASE STUDY: CLIFF PROFILES AT KILVE, WEST SOMERSET

The cliffs at Kilve Beach in west Somerset provide in a microcosm the different ways in which the stratification of rocks determine cliff profiles. Although known as Kilve Beach, there are actually few sandy areas at Kilve, but the coastline is dominated by cliffs on the inland side and an extensive wave-cut or shore platform on the seaward side.

The rocks of the area date from the period of the Triassic, when much of Britain was desert, to the Jurassic, when much of the region was covered in tropical seas. At Kilve, two types of sedimentary rocks are dominant: blue lias, a type of blue-grey limestone of moderate resistance to erosion, and softer beds of shales that vary in colour from black to grey and brown. The limestone is well jointed and has a distinctively blocky structure. The shales are flaky in appearance and are friable to the touch. They are also rich in mineral oil, and on the top of the cliffs on the trackway between the village and the beach are the ruins of a 19th century industrial plant in which the rocks were heated in order to extract oil.

Along much of the cliff face there is alternate bedding of these two materials, each layer varying between 0.25 m and 2 m in thickness; the shales generally form thicker beds than the limestone. The differential erosion of these two materials is clearly visible, having created a stepped face in the cliffs, with the limestone standing out much further than the shale, and this is particularly noticeable towards the foot of the cliffs where marine erosion processes rather than subaerial erosion dominate. Where the shales form the bottom of the cliff faces they produce slopes of less than 45°, whereas the limestone can retain a vertical slope.

These higher cliffs frequently have overhangs and the limestone strata at the top break down into blocks that tumble down to the foot of the cliffs. As the tidal range is high here in the Bristol Channel, the broken off material is quickly processed into rounded boulders and pebbles by attrition and there is clear evidence of the blocks becoming rounded within just a few metres seawards of where they fall. In a few places the higher cliffs have springs sapping through them; at these points there is abundant vegetation and the cliffs are therefore relatively stable.

The highest cliffs at Kilve are around 20–25 m and occur predominantly where the strata are vertical. Where the cliffs have strata that dip either away from the sea or towards it they tend to reach only 5–15 m in height and are noticeably less stable and have stepped profiles. The cliffs with their dip towards the land tend to be higher and their stepped profiles more distinctive than those with their dip towards the sea.

Stretching between the foot of the cliffs and the sea is a wave-cut or shore platorm, which is totally covered at high tide but stretches some 200 m towards the sea at low tide. It dips gently inland at an angle of about 10–15° and forms a series of sloping steps. Where the limestone is exposed there is a well-defined area of tilted limestone pavement, remarkably regular in places. Where the softer shale lies and has been eroded, there are a series of troughs that are now covered by sand and mud deposits.

c) Cliff failure

Cliff failure can be regarded either as the gradual and continuous removal of material from a cliff by the various geomorphological processes at work upon it, or, more commonly, as a sudden and dramatic event that takes away a large amount of material but which occurs only infrequently. The continual toppling, slumping, sliding or flowing processes of mass movement as a result of the daily work of erosion and weathering represent the first interpretation of cliff failure. Events such as the shearing rockfall of 10 January 1999 at Beachy Head in Sussex, which caused the white chalk cliffs to recede by 13 m, represent the second and more dramatic interpretation of cliff failure. The occurrence of cliff failure depends upon geological stability of the cliffs, the climatic regime and the likelihood of freak events that could trigger off a failure on a massive scale. Trigger events could include unusually heavy storms or surges at sea, very heavy rainfall on the land, saturated strata in the cliffs being subject to freezing conditions and earth tremors in regions of tectonic instability.

CASE STUDY: TYPES OF CLIFF FAILURE IN CORNWALL

Although Cornwall's coastline is often dominated by high resistant cliffs, such as the granites of the Penwith Peninsula, the slates of the area around Boscastle and Tintagel and the peridotites, schists and gneiss of the Lizard Peninsula, various forms of cliff failure take place within the county. Many different factors may lead to cliffs losing their original profile and form. These include:

- the interbedding of rocks of different levels of resistance
- the presence of faults or other lines of weakness
- the presence of water in the rocks, particularly where there are both permeable and impermeable rocks in close proximity
- natural joints and cracks in the rocks.

Bristow (1996) recognised five commonly occurring types of cliff failure in Cornwall, which are illustrated in Figure 5. Part (A) shows a stable cliff. Part (B) of the figure shows the instability brought about by the presence of a well-bedded rock, which has its bedding plane dipped towards the sea. Blocks become detached and then slide down slope into the sea. The slates at Ropehaven, near St Austell, exemplify this type of slope failure. Well-pointed rocks, even if they are of highly resistant materials such as granite, may be subject to failure through toppling as shown in part (C) of the figure. The granite cliffs at Land's End fall into this category.

Wedge failure (as shown in part (D)) occurs where two faults or lines of weakness converge and isolate a wedge of cliff material which then comes loose and slips into the sea; there are various small examples of this along the north coast of Cornwall. The occurrence of rotational failure is more limited in Cornwall as it is more associated with softer rocks. This process, shown in part (E) of the diagram, takes place when water in the underlying rocks creates a slip plane and the overlying rocks slide in a rotational way over them. This type of failure is particularly found where small rivers have dissected cliffs and laid down alluvial deposits, or where periglacial deposits form part of the coastline. The type of failure, shown in part (F) of Figure 5, is due to the undermining of the cliffs as a result of the formation of a large sea cave and its subsequent collapse. Lawarnick Pit near Kynance Cove was formed in this way.

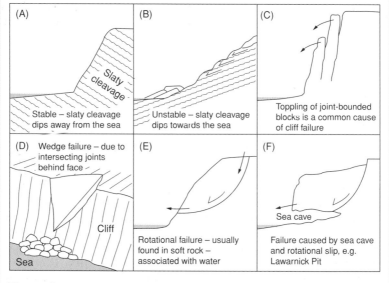

Figure 5 (A) A stable cliff. (B)–(F) Patterns of cliff failure in Cornwall (after Bristow)

3 Coastal Retreat

Rocky coastlines are associated with coastal retreat as they face the full onslaught of the processes of marine erosion, particularly during times of storm. The configuration of rocky shores and the depth of the water that surrounds them encourage the development and breaking of destructive waves. As time passes the shoreline moves back and retreats landwards. Only when the coastlines are very convoluted or the shoreline in front of cliffs is very gently sloping are certain parts of rocky shores protected from the erosion that causes retreat.

There are two main types of shoreline orientation that make coastal retreat take two rather different forms:

- where the cliffs and other features of a rocky coastline lie parallel to the general trend of the shore, the entire coastal stretch is evenly exposed to the erosive power of the sea and therefore the model for coastal recession is one of parallel retreat (this does not necessarily mean that the cliffs keep their same profile as they retreat; high cliffs made of resistant rocks will tend to keep the same profile, whereas cliffs made of less resistant materials may become lower and more rounded in profile as retreat occurs).
- where there is a series of headlands that stick out at right angles to the sea, there is more selective and targeted erosion resulting from the greater force of refracted waves acting upon these headlands, and therefore the model for this type of coastal recession is of the gradual disintegration of the headlands, as shown in Figure 6.

4 Features of Coastal Erosion

As coastal retreat takes place, a whole series of distinctive features may develop within or around the cliffs or flat rocks that form the shoreline. These can be put into categories according to whether they represent the early, middle or later stage in the erosion cycle. Newly formed rocky coastlines or those made of highly resistant materials are likely to be in the earlier stage of development where such features as notches, caves and blowholes are common. More mature rocky coastlines where erosion has had much longer to operate are more likely to have features such as arches, stacks and stumps along them, particularly where there are headlands jutting out into the sea. The oldest coastlines upon which erosion has almost reached its temporal limits are characterised by extensive wave-cut platforms, which protect the cliffs from further erosion.

a) Notches, caves, blowholes and geos

(i) Notches

Notches are grooves that are eroded into cliffs and other rock faces by the various processes of marine erosion. Typically, they are restricted to the area on a rock that lies between the mean high- and low-tide marks. Notches are fundamental to the undermining of cliff faces, which is of major importance in coastal retreat as in places of particular weakness they easily lead on to the formation of caves. Notches are typically smooth and rounded in limestone and chalk

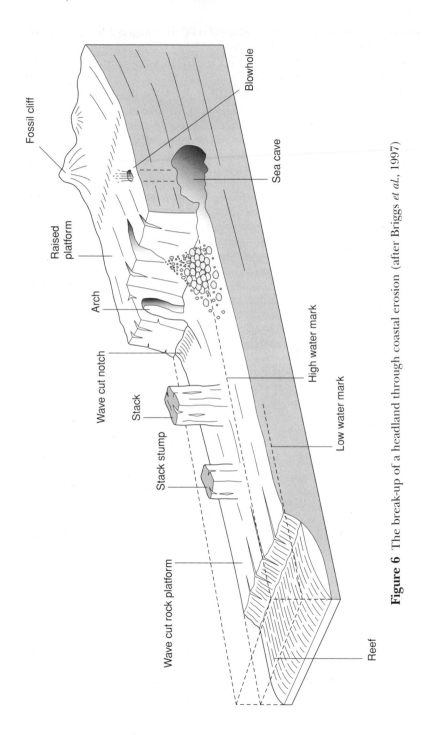

Fossil cliff

Raised platform

Arch

Wave cut notch

Stack

Stack stump

Wave cut rock platform

Blowhole

Sea cave

High water mark

Low water mark

Reef

Figure 6 The break-up of a headland through coastal erosion (after Briggs *et al.*, 1997)

where solution is taking place; this is noticeably so in the Carboniferous limestone at Threecliff Bay on the Gower Peninsula in South Wales and on the chalk cliff base at Botany Bay near Margate, Kent. On tropical limestone cliffs, notches often become deep and cave-like, and have roofs hung with stalactites, as at Liliputa Beach, Samar, in the Philippines

(ii) Caves

Caves may develop from the widening and deepening of notches in places where there are weaknesses in a cliff face. Larger caves may develop where there are bigger places of weakness such as faults and zones of weaker rock embedded in the cliff face, which come under direct attack from the pounding waves. The very large (20-m high) asymmetrical cave in the slate at the entrance to Boscastle harbour in Cornwall mirrors the shape of the fault line in which it lies. In the chalk around Botany Bay, near Margate shallow caves take on a regular trapezoidal form with narrower tops and broader bottoms, and this reflects the relative strength of erosive waves at the very cliff base in comparison with a few metres above it. In limestone areas marine erosion may lead to the exposure of underground caves already formed by solution; these may therefore be much larger than normal sea caves. In tropical limestone areas caves may actually form tunnels right through **towerkarst** islands (islands made from large free-standing pinnacles) and these may be navigable by small craft and be important tourist attractions, as in the Rock Islands of Palau in the Pacific and in Phang Nga Bay near Phuket in Thailand.

(iii) Blowholes

Blowholes are vertical shafts in the cliffs that are linked to the sea, for example, through caves at their lower ends and coming out onto the cliff tops at their upper ends. Blowholes may occur naturally as a result of erosive waves blasting their way through natural lines of weakness in the rocks or where old mine shafts are linked to the sea. The name 'blowhole' relates to the similar feature in whales, through which they expel sea spray. During stormy conditions sea spray comes out of blowholes with great force as plumes of white aerated water. There are several good examples of blowholes in Cornwall, including those on Porth Island at Newquay and the Lion's Den near Lizard Point. In the latter case there is a large rounded open hole that resulted from the collapse of a cave roof. On the cliffs above St Agnes in Cornwall there are blowholes where old tin mine shafts are connected to the sea.

Along the southern coastline of Tongatapu Island in Tonga in the South Pacific, there is a 5-km stretch of low coralline limestone cliffs where there are dozens of blowholes. The Mapu'a à Vaca (meaning the chief's whistles) blowholes shoot some 30 m into the air at high tide.

Blowholes on Tonga, South Pacific

(iv) Geos

Geos are long, narrow gorge-like inlets, which have generally formed as a result of the collapse of the roof of a marine cave. Known as *zawns* in Cornwall, there are numerous small examples in the granite around the Penwith Peninsula. In places, old tin mine shafts have been partially responsible for their formation. On the south Pembrokeshire coast there are also several well-developed examples where there are lines of weakness in the Carboniferous limestone, including the famous Huntsman's Leap.

b) Arches, stacks and stumps

(i) Arches

Sea arches have long been one of the most popular types of coastal scenery with tourists. They are formed by the wearing away of narrow headlands, generally by the formation of two back-to-back cave systems which eventually join. Like all coastal features they are temporary and eventually collapse. Around the British coast there are several spectacular examples of arches, such as the Green Bridge in the Carbonifeous limestone Pembrokeshire and Durdle Door in the Purbeck limestone in Dorset. Close to Durdle Door is the much smaller Bat's Hole, a reminder of what it may have looked like thousands of years ago. Several arches in Britain have collapsed in recent decades, including one in the chalk at Botany Bay in Kent and the free-standing one in the Magnesian limestone at Marsden near South Shields in Tyne and Wear. There was a double arch near Port Campbell in southeast Australia, known as London Bridge, but in 1990 the inner arch collapsed, leaving just the much smaller outer one.

(ii) Stacks

Stacks are tall isolated pillars of rock that are free standing in the sea. They may be alone or occur in groups. They represent either a stage in the break up of a headland or, when they occur parallel to the shore, residual features left behind as cliffs retreat. They may result from the collapse of a sea arch. In Britain, the Needles off the Isle of Wight and the Old Harry rocks off Studland in Dorset are good examples of groups of stacks close to headlands. Similarly, I Faraglioni off the coast of the island of Capri in Italy are a group of stacks belonging to an old headland. Also in the Mediterranean there is a series of stacks at the entrance to the vast natural harbour at Pylos in the Pelopennes, Greece; this was the site of the destruction of the Turkish fleet in the Battle of Navarino Bay in 1827 during the Greek War of Independence. On the coast of Victoria, Australia, the large group of stacks known as the Twelve Apostles were formed as a result of wider cliff recession rather than the break up of just one small headland.

(iii) Stumps

Stumps are small rocky platforms standing offshore, which may be covered at high tide but more commonly are uncovered at all stages in the tide. Undermining of stacks by marine erosion can lead to their collapse and this is one of the main ways in which stumps are formed. Stumps are ideal bases upon which lighthouses can be built, as is the case at the Needles, on the Isle of Wight.

c) Shore platforms

Shore platforms, also known as wave-cut platforms, when they are fully developed represent a late stage in the recession of a rocky coastline as they are what remains after a considerable amount of erosion has taken place and cliff retreat has gone a long way back. Shore platforms are rocky features that extend across the intertidal zone; over time they become both wider and subject to more vertical erosion. A state of equilibrium may be reached beyond which no further downward erosion takes place (unless there is tectonic uplift or some other factor that leads to sea-level change). In the evolution of shore platforms two distinctive types can be recognised:

- **incipient shore platforms**, which are younger geomorphologically and therefore may have many residual upstanding features and irregularities such as small stacks and stumps
- **dissected shore platforms**, which are much older, have been worn flat and have rills and gullies cut into them by streams that run across the platform towards the sea at low tide.

The actual floor of a platform will also depend upon the geology and stratification of the rocks from which it is made. Where stratification is perfectly horizontal, the platform may be very smooth; by contrast, a platform made of sedimentary rocks with dipped bedding is likely to have a stepped floor. Limestone platforms, especially in tropical and subtropical regions, generally develop pitted, honeycombed or pinnacled surfaces as a result of chemical weathering. When less than 10% of a rocky shoreline is flat or smooth it cannot be characterised as a shore platform.

Coastal geomorphologists have put coastlines into four categories in relation to shore platforms. Figure 7 shows these four types of coast-

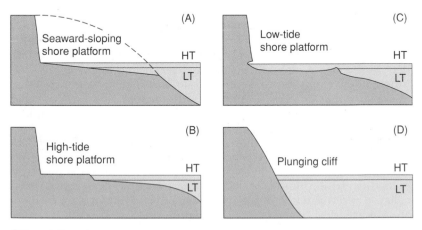

Shore platform types
(A) Seaward-sloping shore platforms
(B) Subhorizontal high-tide shore platforms
(C) Horizontal low-tide shore platforms
(D) Plunging cliffs with no platform on the shore

Figure 7 Shore platforms

lines. In situation (A) the platform is **gently sloping** seawards and has a constant slope, this is typical of places with a high tidal range. In situations (B) and (C) the platforms are of the **quasi-horizontal** type, which develop where the platform is restricted and tidal ranges are generally low. (B) has a marked break in slope at the high-tide end of the platform, whereas (C) has a bench-like feature close to the low-tide mark. Earlier researcher believed that (B) was more typical of the Northern Hemisphere and (C) more typical of the Southern Hemisphere, but this was based upon a relatively small sample of research locations and is no longer shown to be true. However, type (C) is commonly found along tropical and subtropical limestone coasts. The profile shown in (D) has no platform as there are **plunging cliffs** with deep water beneath them. This situation could arise if an old platform has been drowned as a result of sea-level rise.

Summary Diagram

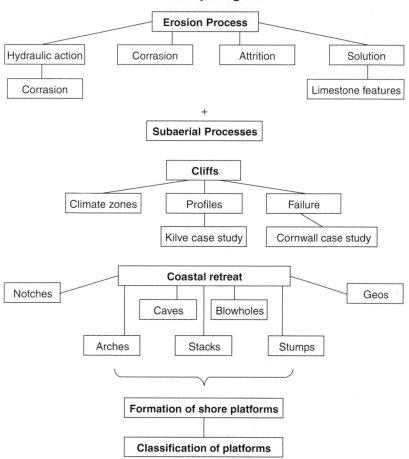

Questions

1. a) Describe the main processes of coastal erosion.
b) Under what conditions may some of these processes be more effective than others?
2. a) With the aid of labelled diagrams explain how caves, blowholes, arches and stacks are formed.
b) Explain what is meant by a shore platform or wave-cut platform and how it evolves.
3. a) Why are cliff profiles so varied?
b) With the aid of diagrams explain how different cliff profiles evolve.
4. a) Why are some cliffs more stable than others?
b) Explain what is meant by cliff failure and under what circumstances it might occur.
5. a) With the aid of diagrams outline the various stages of headland retreat.
b) How does wave refraction influence headland retreat?

3 Coastlines of Deposition and Transportation

When low tides drain the estuary gold
Small intersecting breakers far away
Ripple about a bar of shifting sand
Where centuries ago were waving woods
Where centuries hence, there will be woods again.

North Coast Recollections, John Betjeman

1 The Processes of Deposition and Transportation

The rocky shorelines studied in the last chapter are in marked contrast with the lower lying coastlines typified by beaches and other sediments, where deposition and transportation are the two main processes at work. These are coastlines that are being built up through time rather than being eroded, and they can therefore be regarded as **advancing coastlines** as opposed to retreating ones.

a) Onshore deposition

The coastal deposits that make up beaches and other depositional coastal features are heterogeneous as they are comprised of everything from large boulders, cobbles and pebbles down to the finest clays and silts. Table 1 shows the Udden-Wentworth Scheme, which classifies materials according to their particle size.

Table I The Udden-Wentworth Scheme

mm	ø	Class terms	
		Boulders	
256	−8		
128	−7	Cobbles	
64	−6		
32	−5		
16	−4	Pebbles	
8	−3		
4	−2	Granules	
2	−1		
1	0		very coarse
0.5	1		coarse
0.25	2	Sand	medium
0.125	3		fine
0.125	3		very fine
0.062	4		
0.031	5		coarse
0.016	6	Silt	medium
0.008	7		fine
			very fine
0.004	8		
		Clay	

Beach materials can also be classified into three categories relating to their mode of origin:

- **lithogenic deposits**, i.e. those that have their origins in the breaking down of rocks; this category includes large boulders, cobbles and pebbles
- **mineralogenic deposits**, i.e. those that have their origins in the break up of rocks into certain mineral grains such as quartz and mica; many of these deposits produce material with a sandy texture
- **biogenic deposits**, i.e. those that have their origins in biological organisms such as shells and coral; these materials often break down into a very fine, powdery texture.

Deposits are pushed up onto beaches by constructive waves with their strong swash. During fair-weather conditions the deposits may be limited to sands and silts, but during stormy conditions much larger materials can be forced up onto beaches. Where beaches are backed by cliffs, large pieces of rock that break off the cliff are generally quickly converted into rounded pebbles by the process of attrition.

In order to understand the processes at work in shifting deposits up and down beaches, as well as along them, **sediment** analysis is carried out. This involves the **sampling** and measurement of deposits at various points within a beach profile. On most beaches the coarser material is found on the upper parts that are reached by storm waves but the weak backwash is unable to return it back into the sea. By contrast, the lower parts of the beach have the finer materials, which can be brought back down by the backwash. As pebbles and coarser materials are more permeable than finer ones, they allow the sea-water to pass through them, washing the smaller grained deposits down through. These processes are known as **sediment sorting**. Some beaches which are built up by high-energy waves, such as Chesil Beach in Dorset, are just made up of cobbles and pebbles, and lack smaller sediments altogether. In such places the strong waves that deposit the pebbles onto the beach return any finer sediments that they are carrying back into the sea.

Sediment analysis in Britain has shown that a great deal of beach material comes not from the breaking down of the rocks and pebbled already on the beaches as much as from three other sources:

- cliffs and other rocky areas along the coast that have been broken down by subaerial processes
- sediment supplies brought down into the sea by rivers and then distributed along the coast by currents
- deep sediment sinks that have been lying in the sea since the Pleistocene (e.g. periglacial deposits) which have emerged as a result of sea-level changes and have been deposited on the shore by wave action.

b) Longshore drift

Just as constructive waves are responsible for depositing materials onto the shoreline, waves that are influenced by **prevailing winds** are responsible for moving material along the shoreline. This process, **longshore drift**, takes place where prevailing winds, i.e. those winds which blow most frequently during the year (and are generally also the winds associated most with storm conditions), blow at an oblique angle to the shoreline and thereby cause incoming waves to approach the shore at an oblique angle. Figure 8 shows how the process works. The incoming wave approaches the shore at an oblique angle and its swash therefore also pushes sand, pebbles or other materials up onto the beach diagonally. The backwash from the same wave however, moves back towards the sea at right angles because gravity rather than the wind is the main force at work upon it. As it moves back to the sea, the backwash carries some of the deposited beach material with it. The long-term effects of these alternate swash and backwash

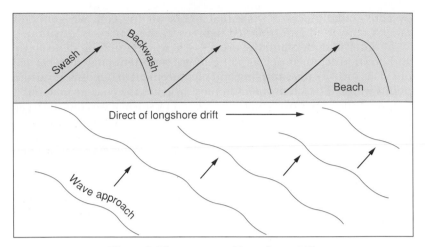

Figure 8 The process of longshore drift

movements are to carry sand, pebbles and other materials along the shore, from one end of the beach to another.

The actual direction that longshore drift takes depends upon both the direction of the prevailing winds and the orientation of the shoreline. In Britain, where the south-westerlies are the prevailing winds, both west- and south-facing coasts come under their influence. Along the south-facing coasts longshore drift transports materials eastwards and along west-facing coasts it transports materials northwards. Although north- and east-facing coasts are protected from the effects of the south-westerlies, they come under the influence of the second most important winds, the north-easterlies. Thus, on east-facing coasts longshore drift transports materials southwards and on north-facing coasts it moves them westwards.

2 The Erosion–Deposition Balance and Beach Compartments

a) Beach budgets

Beach budgets are simply the volumes of sediment supplied to a particular section of beach by onshore waves and longshore drift, balanced out against the sediment lost offshore by backwash movements or onshore by either longshore drift or by aeolian processes. Surveys can be carried out on beach profiles in order to work out how much sediment gain or loss is being experienced over a given period. Calculations can be made to ascertain the rates of longshore drift as well as which sections of the beach have gains or losses. The analysis of sediment grains may also enable coastal geomorphologists to assess the places of origin of materials that are on the move.

Places that are gaining materials and therefore have a positive budget balance are known as **prograded beaches**. Progradation can take place under a variety of circumstances, and these include:

- the shoreward drift of thick sand and shingle accumulations that lie in shallow waters off the coast (as seen in and around Cardigan Bay in Wales)
- along coastlines of emergence, such as those experiencing isostatic recovery since the ice ages (as in the case of the beaches on Holy Island, Northumberland)
- in locations at the ends of spits where longshore drift is bringing new material in (as at the southern end of Orford Ness in Suffolk)
- in places where there are ecological changes taking place that release new sediments (as around the Danish island of Kyholm where the eel grass is dying off)
- in locations where human activity is causing changes in sedimentation (as in the cases of Lowestoft and Yarmouth in East Anglia, where the construction of breakwaters has led to changes in the direction of longshore drift).

Places where there is a negative beach budget through the net loss of sediments both offshore and through longshore drift are simply known as **eroded beaches**. There is a very wide range of reasons why erosion leads to the depletion of beaches; in some cases more than one of these situations may be responsible for the changes taking place. Some of the main circumstances leading to beach erosion are:

- the reduction of sediments being made available from rivers (this frequently happens as a result of upstream damming and a loss of alluvium from floods, e.g. the lack of deposition taking place in the Nile Delta since the building of the Aswan Dam)
- the submergence of the land as a result of relative sea-level changes (this is taking place throughout the Venetian Lagoon and, in particular, along the Lido beaches)
- the reduction of sediment supplies from the sea (this is taking place in many places in Britain that largely became prograded since the ice ages, but in the last century or so the sediment sources have become exhausted, e.g. Newquay in Cornwall)
- the reduction of sediment supplies from eroding and weathering cliffs, especially where human management has led to greater cliff stability, (e.g. Bournemouth, Dorset)
- the increase in wave energy and more stormy conditions increases the erosion levels on beaches, which may be related to possible global warming, (e.g. along the North Sea coast of England in the last few decades which has led to the breaching of the Spurn Head spit in Yorkshire)

- the reduction of the amount of ice cover in areas of high latitude, also related to possible global warming, has led to the coasts being more exposed to erosion, (e.g. in parts of Alaska)
- in a similar way the destruction of coral reefs has left beaches more exposed to erosion, (e.g. on the east coast of Lombok, Indonesia)
- the human removal of materials by dredging (one of the most famous examples of this was at Hallsands in Devon where underwater gravels were extracted in the 1890s to build the Devonport Dockyards; as a consequence both the beach and the fishing village were destroyed by wave action).

b) Beach compartments

Beach compartments, which are also called **sediment cells**, are distinctive areas of coastline clearly separated from other areas by well-defined boundaries such as headlands and stretches of deep water. They may vary in size from a few square kilometres to a few hundred square kilometres, and indeed large compartments are generally subdivided into several smaller ones. Theoretically, within these cells material goes from a **source region** such as a river or eroded cliffs and then undergoes transfer to a **sink** such as deep water or sand dunes to which it is lost. In some cases sediments may get out of the sinks and be recycled, but this process may take many years or even decades. Although in theory beach compartments can be regarded as closed systems from which nothing is gained or lost, in practice it is easy for some finer sediments to find their way around headlands into neighbouring cells.

Large dominating headlands such as the Lizard Point, Dodman Point, Start Point and Portland Bill are all typical major sediment cell boundaries on the south coast of England, but between them there are many much smaller compartments. For the purpose of coastal management the coasts of England and Wales are subdivided into 11 'Littoral Compartments', such as that stretching from Land's End to Portland Bill – these take into account the nature of sediment cells.

3 Beach Profiles

Beaches have many distinctive features both in their cross sections and in their aerial plans. These features reflect the changing patterns of swash and backwash, the differences between summer and winter seasons, the incidence of storms and the way in which the working of beach compartments create either a negative or a positive sediment budget.

Figure 9 is a three-dimensional diagram of a beach from which the main features of its profile and plan can be clearly seen. It is an

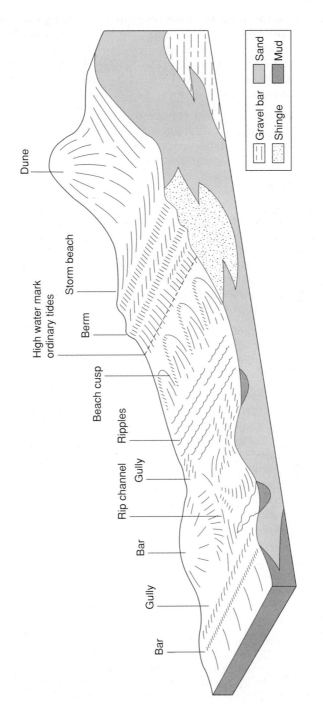

Figure 9 An idealised beach profile (after Briggs *et al.* 1997)

idealised profile of a complex beach composed of both sand and shingle, and has been given the full range of features that occur on different beaches. The upper beach is dominated by wave-imposed features, is made of coarser materials than the lower beach (e.g. pebbles, gravels and coarse sands) and maintains an angle of around 10–20°. The main features of the upper beach are:

- the **storm beach**, which is a very distinctive ridge marking the level to which pebbles and other materials are pushed during the very highest spring tides, particularly during storms; it may remain in place for many months, until other storms change its position
- **beach ridges**, which are also known as **berms**, are other lower ridges parallel to the storm beach that mark the successive levels to which materials have been pushed during earlier storms and high tides
- **beach cusps**, which are crescent-shaped abayments cut into the shingle ridges that may vary in size from 1 m across to as much as 30 m; there are many different theories as to their origins and one of the more widely accepted is that they are formed by two converging streams of swash that scoop out arc shaped slopes in the shingle.

The lower beach has a much gentler gradient of down to as little as 1–2° and is made of fine materials such as fine sand and silt. The main physical features of the lower beach are:

- small and often heavily braided **drainage channels** cut into the sand through which water from the swash drains back downslope
- **ripples**, which develop over the vast exposed areas of sand flats, either as a result of tide action or wave action (if the sand flats have been dried out for any length of time wind action may also be responsible for ripple formation)
- low-lying, flat topped **sand ridges**, sometimes called **intertidal bars**, which develop parallel to the shoreline in places where breaking waves are able to deposit fine materials; the water of the outgoing tides also help to shape them
- **runnels** which are the shallow depressions which form between the ridges and are therefore also parallel to the shore; these are filled with shallow temporary lagoons, which change in depth with the state of the tides.

There is also generally at least one further ridge of sand underwater beyond the low-tide mark, known as a **subtidal bar**. It is this feature that causes people wading out into the sea to experience the water becoming shallower a few metres out. Subtidal bars are rarely continuous and are generally dissected by **rip currents**, which are fast moving streams of backwash that carve out **rip channels** between them. Rip currents may also cut through intertidal bars.

4 Sand Dunes

As coastal environments tend to be exposed and windy places, aeolian processes are important in the development of landforms where there are fine sediments such as sand and silt. As the wind blows across the drier sand deposits on a beach it takes material inland, where it starts to accumulate as sand dunes. The optimum conditions for the formation of sand dunes are as follows:

- there is sufficient flat land available to accommodate the dunes
- there are strong onshore winds
- there are plentiful supplies of sand with the right grain size
- there is vegetation to colonise the dunes (this is not necessarily the case along arid coastlines)
- that the beach ideally has a low gradient and a high tidal range in order to provide a large sandy surface area over which the wind can blow.

a) Sand movement

The sand that makes up dunes moves in a number of different ways. Where there are small and damp grains that tend to stick together and therefore form particles larger than sand grains, they move across the beach by **surface creep**, a slow, rolling motion. Most dry sand grains move by a series of jerks and leaps, a process known as **saltation**. Once many grains are set in motion, falling grains impact with and dislodge other grains. In this way less energy is needed for the grains to move; this situation is called the **impact threshold velocity**. During very high velocity winds, sandstorms may occur and this leads to sand grains being held in the wind and carried long distances by the process of **suspension**.

b) Sand dune forms

Coastal sand dunes take a variety of forms, according to both their position on the beach and the shape that they take. They do, however, lack the great variety of forms found in desert areas.

At the most simple level, sandy beaches are generally covered in thousands of small ripples a few millimetres in height, which are effectively minute dunes. The first larger dunes to establish themselves at the top end of the beach are known as **embryo dunes** in which the pioneer plant species take root. Beyond these the dunes form into parallel ridges, with the first ridges known as **fore-dunes** and those further back as **hind-dunes**. Small crescent-shaped **parabolic** dunes a few centimetres high may form in the lee of clumps of grass and small bushes. Much larger parabolic dunes form where strong winds and other factors such as human trampling give rise to dune collapse; these are known as **blow outs**. Within sand dune complexes there are generally areas where the water table reaches the surface and there are water- or marsh-filled hollows known as **dune slacks**. Old dunes that have become highly stabilised, have thick vegetation growing on them and may have developed a distinctive soil profile are referred to as **remanié dunes**. Some old dunes may even form into a much harder, cemented material that can be quarried and used as a building material, as it is in many parts of the Mediterranean; these are therefore known as **lithified dunes**.

The ecology and ecosystems associated with sand dunes will be examined in the next chapter.

5 Other Features of Coastal Deposition

In addition to the more ubiquitous sand or shingle beaches and sand dunes, there are several other important features of coastal deposition that are more restricted in their occurrence and which develop only in specific locations when the conditions are right.

a) Bars and barrier islands

Bars or barriers are depositional features laid down in the sea parallel to the shoreline by constructive waves and are separated from the shore by lagoons. They may vary in size from a few hundred metres long to a few thousand, and may be from tens of metres to hundreds of metres in width; the lagoons may also vary in width from a few hundred metres to tens of kilometres. Barriers develop typically in seas with shallow gradients and with low tidal ranges.

Within Europe the biggest system of barrier islands are the Friesian Islands, a chain of 11 larger islands and many more smaller ones, which stretch round from the northern part of the Netherlands into Germany; they enclose a large area of shallow sea known in Holland as the Waddenzee. The Venetian lagoon in Italy is also bounded by a series of long, narrow barrier islands (and this will be looked at in more detail in the last chapter).

There are numerous barrier islands along the West African coast where wave action is strong but tidal ranges are limited. The main concentrations of barriers are in the Ivory Coast, Benin and Nigeria. Lagos, the former Nigerian capital, takes its name (Portuguese for 'lakes') from its coastal lagoons. In some cases the islands are totally separated from the mainland, in others they have migrated and joined onto the shore, at either one end or both.

The biggest concentration of barrier islands is in the USA where they can be found offshore along many parts of the east coast from Long Island, New York state, through New Jersey, Delaware, Maryland, the Carolinas, Florida, Louisiana and along the coast of Texas to the Mexican border. The most spectacular formations are to be found along the more than 500 km of barrier islands, which converge at the cusp-shaped Cape Hatteras in North Carolina and enclose the huge lagoon of Pamlico Sound.

There are four basic theories of how barriers and barrier islands have formed; each of these is equally valid as in some places one mode of origin may provide the explanation whereas in another place an alternative explanation would be more relevant. The four theories are:

- during a period of stationary sea levels there is a build up of long-shore bars from storm swash and blown sand accumulation
- when sea levels are rising the shore is inundated and storm beaches and sand dune ridges are isolated to form offshore barriers, which are further added to by marine and aeolian processes; these are known as **regressive barriers**
- when sea levels are falling **transgressive barriers** form; this involves underwater longshore bars and spits being exposed and migrating inland through both wave and wind action
- when spits are heavily built up by storm waves and then breached, a series of islands are formed with narrow stretches of water between them.

b) Spits

Longshore drift moves sand, pebbles and other sediments along shorelines. When there is a major change in the trend of the coastline, such as where it opens out into a broad bay or estuary, the longshore drift may continue to deposit sediments into the sea, creating an elongated beach that projects out into the bay or estuary. This feature is a **spit**. Spits may grow at fairly rapid rates; Orford Ness in Suffolk, for example, has been found to be extending southwards at a rate of up to 15 m per year.

Many spits have recurved or hooked ends that may result from wave refraction, currents or a combination of the two. In certain cases, such as at Blakeney Point in north Norfolk, spits may have several recurves, and these date from different stages in their development. Spits often create sheltered areas behind them and frequently this enables the formation of salt marshes away from the force of the swash of incoming waves. Salt marshes are well developed behind such spits as those at Dawlish Warren in Devon and Hurst Castle in Hampshire.

Spits are capable of developing to such an extent that they can cut off large sea areas and convert them to lagoons. The 98-km long Neringa Spit in the Baltic Sea (partially in Lithuania, partially in the Kaliningrad Oblast of Russia) has virtually sealed off the lagoon behind it but the authorities keep a large channel dredged in order to leave shipping lanes to the coastal port of Klaipéda open.

There are some 30 major spits around the coasts of England and Wales alone, and there are examples of which that are growing in each of the four cardinal directions according to their positions in relation to the prevailing winds (although the general direction of growth is that stated below, local factors such as recurving may mean that the orientation of the spit deviates somewhat from each cardinal point):

- northwards: Ro Wen and Morfa Dyffryn (across estuaries in Cardigan Bay, Wales)
- southwards: Orford Ness, Suffolk and Spurn Head, Yorkshire
- eastwards: Dawlish Warren, Devon and Hurst Castle, Hampshire
- westwards: Blakeney Point and Scolt Head (Island), Norfolk.

In some places there are **double spits**. These occur within a given sea area such as a bay, where there are two opposing swells and therefore two spits developing in converging directions. This occurs at Poole Harbour in Dorset where the Studland Bay spit is developing eastwards and the Sandbanks spit is growing southwards and there is only a narrow stretch of sea between them. If two converging spits meet they would create a cuspate foreland.

CASE STUDY: DUNGENESS, KENT

Seen from the air, Dungeness has an elegantly sculpted cuspate form (see Figure 10). At ground level, however, its extensive open landscape composed of numerous shingle ridges, known as 'fulls', is interrupted by the ugly shapes of nuclear power stations, gravel extraction plants and holiday homes. The foreland contains 40% of Britain's coastal shingle and its mode of formation has been a matter of great debate in the past. The most widely accepted theory of its evolution has been the changing positions of the seaward edge of the main shingle ridge over roughly a 3000-year period.

The spit started off with shingle eroded from the eroded cliffs at Fairlight in Sussex, pushed north-eastwards by the prevailing winds (stages 1 and 2). Eventually the spit extended the whole way across the bay to Hythe (stage 3). A rise in sea level then led to more raid erosion of the cliffs to the south-west, which in turn left the shingle more exposed to the prevailing winds causing the spit to swing around to a west–east alignment through longshore drift (stages 4 and 5). The biggest build up of material is at the nose of the ness, where wave refraction causes a loss of energy. Also, north-easterly winds have helped to shape the eastern shore of Dungeness (as in stage 6).

Today, the southern shore of Dungeness is still experiencing the deposition of shingle, whereas the east coast is being eroded and sea walls are necessary to protect it.

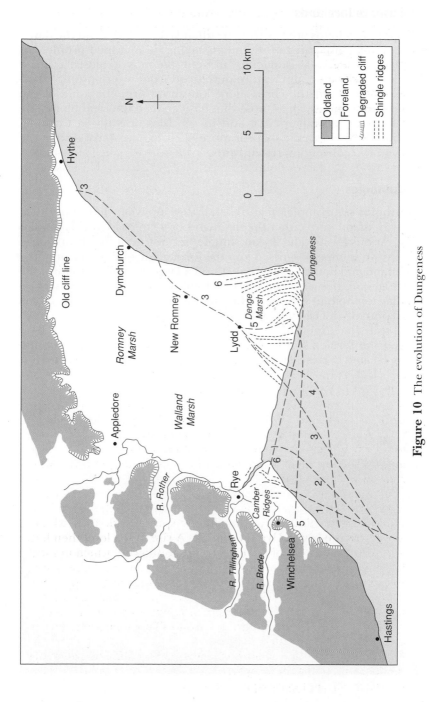

Figure 10 The evolution of Dungeness

c) Cuspate forelands

Cuspate forelands are low-lying headlands that result from the development of a large area of sediments taking on a cusped profile. On the larger scale, as at Dungeness in Kent, the most extensive cuspate foreland in Europe, the feature developed from two opposing directions of swell and the consequent longshore drift. In some of the smaller examples, such as those that have created the tombolos along the Western Australian coast and in the Isles of Scilly in Cornwall (mentioned below), the shape is more likely to have derived from the work of two converging rip currents.

d) Tombolos

Tombolos are complex features that develop when longshore drift joins an island onto either the mainland or a larger island. Tombolos take various forms. The 28-km shingle bar of Chesil Beach in Dorset has grown south-eastwards to join the former Isle of Purbeck onto the mainland and in the process has cut off a large area of sea to form the Fleet lagoon; this is Britain's largest tombolo. Perhaps the most spectacular tombolo in Europe is the one that gave its name to all others the Tombolo di Orbetello in Tuscany, Italy. In this case there are two sand spits (one trending eastwards the other northwards) which have joined the former Monte Argentario island onto the mainland and have created a large lagoon within which the town of Orbetello is

Toll's Island tombolo, St Mary's, Isles of Scilly

located on a causeway. The lagoon was also the location of Italy's fleet of flying boats during the Fascist era of the 1930s and 1940s.

Many smaller tombolos have formed where cuspate forelands link up with islands. There are many examples of this along the coast of Western Australia where old patches of coral reef have been attached to the mainland. In the Isles of Scilly off Cornwall, there are several examples of these smaller tombolos, including that which links Toll's Island to St Mary's.

6 Estuaries and River Mouths

An estuary can simply be defined as the last section of a river before it flows out into the sea, where there are tidal influences and therefore daily fluctuations in the water levels. Estuaries, like deltas, are zones where the marine processes and fluvial processes are both of great importance and the nature of each individual estuary is determined by which processes are dominant. Within a typical estuary there are four main zones according to the levels of salinity and sediment type:

- the estuary head where salinity is between 0.5 and 5 ppt (parts per thousand) and muds are the main sediment type
- the upper reaches where the salinity is 5–18 ppt, and muds are still the dominant sediment type
- the middle reaches where the salinity is 18–25 ppt and there are both muds and sands
- the lower reaches where the salinity is 25–30 ppt and sand is the main sediment type.

Beyond these four zones is the actual point where the river flows into the sea, the river mouth. Here deposits may continue to be deposited into the sea and, under certain conditions, deltas may form. According to Haslett (2000), river mouths where deltas are being formed fall into three broad categories.

- Inertia-dominated mouths, which occur where there is a fairly steep drop off in coastal slope that enables the river's load being carried out into the sea to spread both vertically and laterally.
- Friction-dominated mouths, which occur where the coastal waters are very shallow and the deposition can only spread laterally at the river mouth.
- Buoyancy-dominated mouths, which occur where the river flow into the sea is like a straight jet of water that does not allow lateral extensions of the deposits along the coast.

Estuaries have the same influences working upon them as deltas; they are affected by rivers, waves and tides, they are areas of great accumulations of sediment, often referred to as **sediment sinks** and are relatively young features in geological time and are constantly developing.

Estuaries are characterised by different types of water movement within them and it is the relationships between sea water and freshwater within them which led the US marine scientist Prichard to draw up his classification of estuaries in the 1950s. Prichard recognised three types of estuaries.

- Stratified estuaries, which are dominated by the flow of the river and have very little turbulence or current influences. These estuaries are characterised by layers of different types of water, with the freshwater on the top. The Hudson River in New York has this form of estuary.
- Partially mixed estuaries, which have the river dominating the water in some parts of the estuary, but tidal water dominating other parts The huge area of Chesapeake Bay in Maryland and Virginia into which the Potomac, James and other smaller rivers all flow, is an example of this type of estuary.
- Mixed estuaries, which have a complete mixing of freshwater and salt water so that the layering of salinity within them is vertical rather than horizontal. These tend to occur where the waves or the tides have a stronger influence than the flow of river water. An example of this type of estuary is the Bay of Fundy in Canada, which has the highest tidal range in the world.

The Mawddach estuary and the Barmouth viaduct

7 Tidal Flats

Around the edges of estuaries there are generally extensive unvegetated depositional areas known as **tidal flats**. These areas are intertidal, as they are covered up during high tides and exposed during low tides. Intricate patterns of channels and rills that dissect them are visible at low tide as the only parts of the estuary to be filled with running water. The individual tidal flats may be composed just of mud or sand, but more often they are a mixture of the two. In the estuaries of the Afon Mawddach near Barmouth and the Afon Dyfi at Aberdyfi on the west coast of Wales, the lower parts of the tidal flats are dominated by sandy deposits brought in from the sea and the upper reaches are more dominated by mud flats deposited by the rivers.

Tidal flats can become quite complex interbedded deposits as a result of both changing tide patterns and occasional river flooding. Cross sections cut into these flats can therefore reveal recent historical changes in flow patterns of both rivers and tides, as well as the effects of major storms over both the land and the sea.

Tidal flats also provide ideal environments for certain organisms such as worms and bivalves to burrow in order to feed and take protection. These creatures churn up the sediments by their burrowing, a process called **bioturbation**. There may be dozens of worms such as the lugworm, a favourite source of bait for anglers, per square metre and therefore bioturbation may completely obscure the interbedding, which occurs within the upper layers of tidal flats.

Summary Diagram

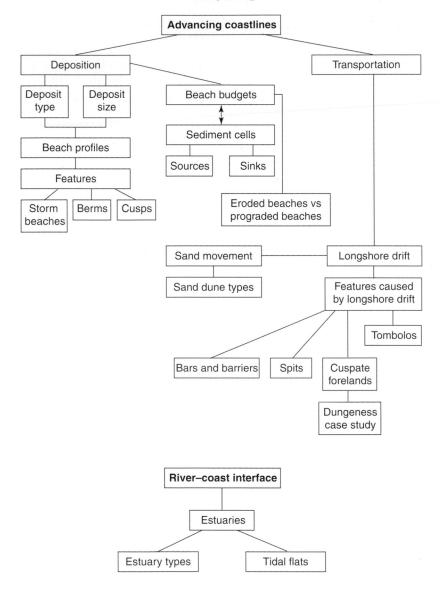

Questions

1. **a)** Draw a labelled diagram of an idealised beach profile.
   b) Explain how the various features you have labelled evolve upon beaches.
2. **a)** Describe the following features: spits, barrier beaches, tombolos and cuspate forelands.
   b) For each of them, with the help of labelled diagrams, suggest their mode of origin.
3. **a)** What are the main features to be found in estuaries?
   b) Explain the different processes that give estuaries their distinctive features.
4. **a)** Define the terms: sediment sorting, sediment analysis and sediment cells.
   b) Why do coastlines of deposition vary so much in character?
5. **a)** Outline why constructive and destructive waves create such contrasting coastlines.
   b) For any three features of a lowland coastline explain the circumstances under which they develop.

4 Coastal Ecosystems

The deeps have music soft and low
When winds awake the airy spry,
It lures me, lures me on to go
And see the land where corals lie.
The land, the land where corals lie.

Where Corals Lie, Richard Garnett

1 The Range of Coastal Ecosystems

The sea and its shorelines create the conditions in which a whole series of different and diverse biogeographical environments may develop. The most important factors that influence which plant species grow where include:

- the nature of the shoreline and its parent material (whether it is rocky, sandy or muddy)
- the climatic conditions of the specific area (especially temperature ranges and rainfall totals)
- the salinity of the sea water (whether it is brackish, normal or concentrated)
- the tidal range of the sea
- the frequency of storms and rough seas.

Some coastal environments can be found in most parts of the world, e.g. sand dunes and rocky shores, whereas others are restricted to tropical and subtropical climate belts, e.g. mangrove swamps and coral reefs.

Although all coastal ecosystems that are related to specific local conditions can be put into two broad subdivisions of **hydroseres** (those associated with waterlogged and wet environments) and

xeroseres (those associated with freely drained or dry environments), it is important to subdivide them into other types of seres. Along the coasts can be found three distinctive types of seres:

- **psammoseres** or sand dune ecosystems
- **haloseres** or salt-tolerant ecosystems, which include **salt marshes** and the **sebkhas** or salt pans found in coastal areas in arid lands, as well as tropical **mangrove swamps**
- **lithoseres**, which are those associated with bare rock surfaces or broken down rock fragments, such as pebble beaches and mass movement deposits.

In addition to these are **coral** ecosystems that cannot be normally classified as a type of sere as the coral polyps are a form of animal life rather than plant life.

2 Psammoseres

a) The formation and structure of psammoseres

The previous chapter looked at the formation of coastal sand dunes as a result of both wave action and aeolian processes, together with the way in which dunes move. This information is important in understanding the type of botanical activity that takes place in coastal dune systems. It is the very mobility of the sand that represents the first obstacle to plant colonisation. Once certain pioneering plant species have taken root, dune stabilisation starts to take place and the whole sequence of plant types that make up the psammosere can begin. As psammoseres can be found in many different climatic zones of the world from the tropics through the subtropics to the cool and even cold temperate parts of the world, the plants that colonise the dunes may vary considerably from one psammosere to another. Nevertheless, they do have things in common. A large proportion of the plants have to be **xerophytes**, adapted to dry conditions, both because of the lack of water retention in sand dunes and the frequency of winds that have a drying effect upon coastal plants. In addition to these causes of aridity, high summer temperatures within the dune systems (especially where the dunes are dark in colour) exacerbates the situation.

At the same time many of the species are **halophytes** or plants adapted to salty conditions; this is especially true of those species that take root early on in the psammosere sequence in positions close to the sea where onshore winds and sea spray both bring salt particles inland. The formation of sand dunes and the movement of sand grains were dealt with briefly in the last chapter, and it is within that context that that the role of plant activity should be examined.

There are many well-developed psammoseres around the coast-lines of Britain and these include: Blakeney Point and Scolt Head in

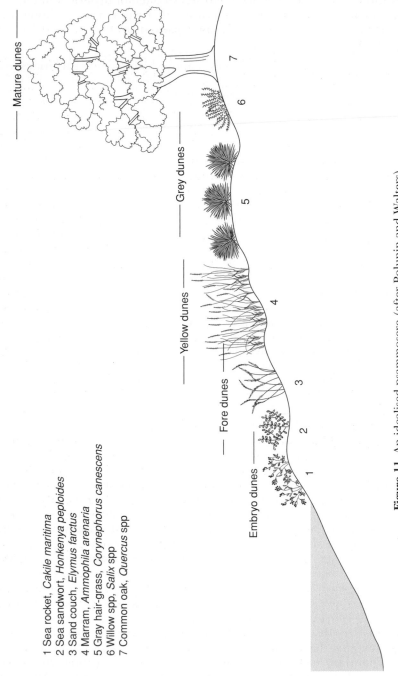

1 Sea rocket, *Cakile maritima*
2 Sea sandwort, *Honkenya peploides*
3 Sand couch, *Elymus farctus*
4 Marram, *Ammophila arenaria*
5 Gray hair-grass, *Corynephorus canescens*
6 Willow spp, *Salix* spp
7 Common oak, *Quercus* spp

Figure 11 An idealised psammosere (after Polunin and Walters)

Norfolk, Braunton Burrows in North Devon, Morfa Harlech in Wales and Culbin Sands in Moray, Scotland.

Figure 11 represents an idealised psammosere of the sort found around the British coast.

On the parts of the beach known as the upper strand plants do not take root because the area is often covered by sea water, especially during high tides and storms. The sand is dry for long enough periods, however, for some particles to be picked up by the wind and to be blown further in land. It is this that forms the **embryo dunes**, which are small and low, and they are the first dunes to be colonised by plants. The pioneers tend to be succulent xerophytes such as sea rocket (*Cakile maritime*) and saltwort (*Salsola kali*). Once these take root, they start to stabilise the dunes and to create some humus.

The next range of dunes moving inland are the lower ridges or **foredunes**, which are colonised by grasses such as sea couch grass (*Agropyron junceum*) and lymegrass (*Elymus arenaris*). Next come the higher foredunes, which are highly mobile and often referred to as **yellow dunes**, as their colour is not greatly altered by the presence of vegetation. It is here that, in addition to the other species, marram grass (*Amophilia arenaria*) becomes the dominant plant and dune stabiliser. Marram grass has long rhizome and root systems that hold the sand dune material together. Once established, it enables many other plants to colonise. In the middle stages of dune fixation, three groups of species move in, all of them contributing to the development of humus and therefore a rudimentary topsoil; these are:

• ephemerals, such as wall saxifrage (*Saxifraga tridactylites*)
• perennial herbs, such as ragwort (*Senecio jacobia*)
• mosses, lichens and liverworts.

Further inland are the hind dunes, which are fixed and often referred to as **grey dunes** because their original sandy colour has been transformed by the high humus content from the dying vegetation. In these dunes there is a much greater range of biodiversity and a large proportion of the species are plants that grow equally well in non-marine environments. In this later stage of development there are herbs, shrubs and trees all present. Not far inland from these dunes, normal climax vegetation of willow, oak or birchwoods may be found.

The sequence of dunes is often interrupted by the appearance of **slacks** or troughs in between the dune ridges where the water table comes close to the surface. In these places the plant species are once again adapted to the particular conditions. There are two types of slacks according to the average level and seasonal changes in the water table: **dry slacks** where plants such as sea holly (*Riyngium maritimum*) and sea spurge (*Euphorbia paralias*) colonise and **wet slacks** where plants such as creeping willow (*Salix rapens*) and bog myrtle (*Myrica gale*) thrive.

The other major disruption to the normal psammosere sequence is the occurrence of **blow outs** where an individual dune has

A sand dune blow out, north Norfolk

collapsed, creating a crater like hollow. There are various reasons why this occurs, for example heavy storms, the incursion of the sea leading to erosion, large rabbit colonies and human interference. Whatever the cause, the formation of a blow out leads to the destruction of the biodiversity within its vicinity.

b) The degradation of coastal dunes

Sand dunes have the appearance of being hardy, but this belies the fact that they are really fragile ecosystems. The difficulties that plants have in colonising dunes in the first place, the liability of blow outs occurring and the activities of human beings all contribute to the susceptibility of dunes to degradation.

When blow outs occur, up to 50% of the sand in a dune may be lost and it may take decades for vegetation to re-establish itself. Once a blow out has taken place, a dune is exposed to further erosion and degradation. Certain more delicate dune plants may be unable to survive within the blow out hollow, and habitats for birds, amphibians and reptiles may also be destroyed. These habitats may have rare species of toads and sand lizards. By contrast, it is the burrowing by rapidly breeding rabbit populations that often provokes or speeds up the rate of dune collapse.

Human activity is particularly problematic. Dunes are seen as ideal places for a wide variety of types of recreation. Huge areas have already been lost to golf, a sport that originated in coastal dunes. Other pursuits, whether they be more gentle like just walking in the dunes, or more active and high speed such as motorcross, mountain biking and driving dune buggies, are all detrimental to the natural dune environment. Perhaps the greatest volume of destruction is merely that done by summer visitors taking short cuts from car parks to beaches, thereby causing large amounts of dune trampling

Action taken to safeguard sand dunes from degradation can take a number of forms, including:

- the cordoning off of areas of dunes that are particularly sensitive and restricting public access to them
- the transforming of importantly biodiverse stretches of dunes into nature reserves with access limited to certain routes, and many information boards to educate the general public about the importance of the dune ecosystems
- the building of boardwalks through the dunes to be used rather than informal pathways
- the stabilisation of dunes by planting successful species on dune ridges and by creating shelter belts to protect dunes from blow outs.

3 Haloseres

a) Salt marshes

Salt marshes form in sheltered spots along coastlines, such as where a spit or a sand bar protects low-lying mudbanks from wave attack. The initial impetus for the development of a salt marsh comes when sea plants such as eel grass (*Zostera marina*) trap sediments above the low-water mark and build up small terraces of mud and matted vegetation. This allows further accretion to take place. At the same time the process of **flocculation** is taking place; this involves the clay particles in the sea or an estuary coming into contact with salt particles, causing the clay to coagulate into larger particles known as **flocs**. As these flocs become heavier, they add to the accumulated deposits. Some of the faster growing salt marshes in Essex experience a vertical accumulation of around 15 mm per year.

Figure 12 shows the typical sequence of plants found in a salt marsh halosere in Britain. The first truly land plants to colonise salt marshes are of the glasswort family, such as the European glasswort (*Salicornia europea*). These can tolerate many hours of submersion per day, and their stems trap mud and add to the accumulation of sediments that enable the next series of plants to colonise. The salt marsh grass (*Puccinella maritime*) is the next dominant species in the sequence; it can tolerate some daily submergence and is one of the fastest salt marsh builders, capable of accumulating up to 5 cm of silt per year. Above the grasses zone are the rushes including the salt marsh rush (*Juncus gerardii*), which tolerate short periods of water cover. On the upper zone, which is only covered in water at storm or

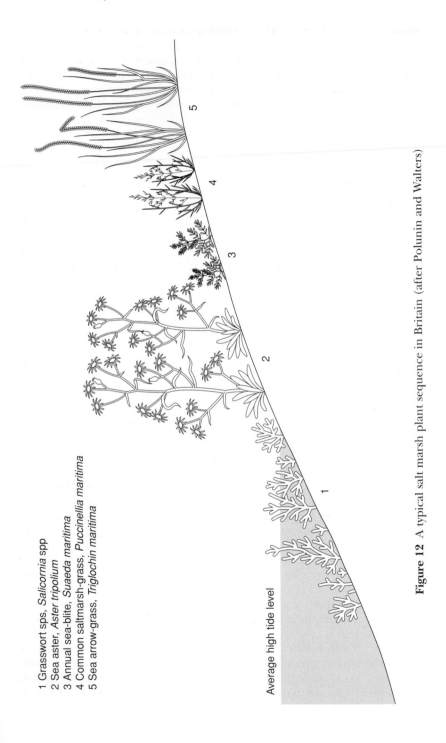

1 Grasswort sps, *Salicornia* spp
2 Sea aster, *Aster tripolium*
3 Annual sea-blite, *Suaeda maritima*
4 Common saltmarsh-grass, *Puccinellia maritima*
5 Sea arrow-grass, *Triglochin maritima*

Average high tide level

Figure 12 A typical salt marsh plant sequence in Britain (after Polunin and Walters)

flood time, a variety of plants flourish such as the sea rush (*Juncus maritimus*) and wild celery (*Apium graveolens*). These plants lie between 20 and 40 cm above the high-tide mark. Between the terraces are hollows containing pools and ditches, creeks and salt pans, and there are variations in vegetation accordingly. For example, in the pools and ditches plant communities include the beaked tassleweed (*Rupia maritime*) and the holly-leafed naiad (*Najas marina*).

Salt marshes, like any other ecosystems, change dramatically when a new, aggressive species is introduced. Many British salt marshes and tidal flats have suffered as a result of the accidental introduction of spartina or cord grass from around the 1870s. Various versions of spartina such as Townsend's cord grass (*Spartina townsendii*) and the English hybrid, common cord grass (*Spartina anglia*) colonised many of Britain's main salt marshes and tidal flats with a vengeance in the decades since their introduction and are widely viewed as a problem. Examples of coastal flats where spartina grass has rapidly taken hold include the Sand Bay area, to the north of Weston-super-Mare in Somerset, and the north side of the Kent estuary in Cumbria near Kent's Bank and Grange-over-Sands.

The consequences of spartina introduction include:

- a rapid spread in the actual area of coastal Britain covered by salt marshes
- the transformation of many previously uncolonised mudflats into salt marshes
- the exclusion of native salt marsh species as a result of aggressive spartina colonisation
- as spartina dies back after about 25 years, the areas where it is well established suffer from salt marsh erosion once it does die back.

In addition to the invasion of alien species, the biggest threat of all to the traditional, natural salt marshes is human activity. Salt marshes and other coastal flats are ideal places for land reclamation for farming, urban expansion, and the development of heavy industry and power plants. Although urban and industrial development can obliterate former salt marshes, agriculture may leave them green – but far removed from their natural state.

CASE STUDY: SALT MARSHES IN THE BRIDGWATER BAY AREA, SOMERSET

There are extensive areas of both salt marshes and reclaimed salt marshes along the edge of Bridgwater Bay, part of the Bristol Channel off the coast of Somerset. The mudflats, and the salt marshes that colonised them, developed in the shelter from the sea provided by coastal dunes and the shingle spit of Stert Point, both the result of strong westerly winds and longshore drift. Beyond Stert Point is the low-lying Stert Island, also the result of the accumulation of sediments in Bridgwater Bay.

Much of the zone to the south of the spit is a nature conservation area where the salt marshes have been little affected by human activity; this is indeed the largest area of salt marsh in Somerset. There are four identifiable botanic zones in the Stert salt marshes:

- at the lowest tideward level spartina or common cord grass has completely taken over from its predecessors and is now the primary coloniser
- slightly higher up are a community of plants that include salt marsh grass and sea aster (*Aster tripolium*)
- the common reed (Phragmites australis) dominates the next level up
- at the highest landward side of the salt marsh, some of the land is grazed and some left untouched; sea couch and sea club rush (*Scirpus maritimus*) are common in the grazed areas, but plants such as sea wormwood (*Artemesia maritima*) and common scurvy grass (*Cochlearia officianalis*) are also important members of this community

Several rare plant species are to be found on Stert Island, protected from the outside world by its isolation and restricted access.

Most of the other nearby marshes were reclaimed from the sea to produce pasture land during the 18th and 19th centuries; these include Steart [*sic.*] Moor, Pawlett Hams and Wick Moor. Today, they are mainly improved pasture for dairy cattle, which is dissected by a network of *rhymes* or drainage ditches. Most of these ditches have fresh water in them, but in places there are stretches of brackish water that support halophytic vegetation associated with coasts, such as the sea clubrush (*Scirpus maritimus*) and the brackish water crowfoot (*Ranunculus baudoti*).

Within these different types of wetland, there are over 22 rare insect species supported by the vegetation. These environments are also important feeding grounds for wading birds and wildfowl. Such factors show why it is so essential to maintain these coastal ecosystems.

b) Sebkhas

Sebkhas or **sabkhas** are the arid zone equivalents of temperate salt marshes. Typically, they are lagoon and salt pan complexes cut off from the sea by a barrier beach or row of dunes. Sebkhas are common along the southern shores of the Gulf in countries such as Bahrain and the UAE, and are also found extensively in North African countries such as Tunisia and Algeria, along the Mediterranean shores. High tides and strong winds, as well as occasional heavy desert rainstorms, lead to the sporadic flooding of sebkhas. Within a sebkha area there are likely to be three main zones:

- immediately inland from the coastal barrier a lagoon area that will vary in salinity according to the amount of fresh water seeping out of springs, the season of the year and the amount of sea water that percolates through the dune or barrier; where there is a large amount of fresh water involved, the lagoon may support fairly dense vegetation including reeds, sedges and aquatic grasses
- at the furthest point inland are salt-encrusted flats covered in evaporite deposits such as halite, gypsum and anhydrite which prevent any form of plant colonisation
- between the lagoon and the salt flats is a transitional zone where salt-tolerant forms of algae may thrive and create deposits of 'algal mats'.

c) Mangrove swamps

(i) The formation and occurrence of mangroves

Mangroves are a range of tree and bush species that are adapted to life in coastal swamps and muddy creeks and estuaries in tropical waters. Water is brackish enough in some larger estuaries to enable mangroves to grow up to 100 km inland. Although mainly in tropical areas, they also extend into the subtropics and are found as far north as 32°N on the island of Bermuda and as far south as 38°S on the North Island of New Zealand. Mangroves often grow in dense, closed and impenetrable thickets and are located mainly between the midtide and the high-tide marks; their pioneer species, however, grow close to the low-tide mark.

Worldwide, there are hundreds of different species of mangrove trees and bushes, and each area where they grow has a distinct formation or association of plants. For example, within the mangrove swamps of Queensland, Australia, there are over 50 different species, some of which are found in other parts of the world, others of which are unique to that region. To the untrained eye the plants within a mangrove swamp may look monotonously similar, but on close examination there are big variations in leaf shape, the type of fruit and the methods of seed dispersal. The biggest variations between mangroves, however, are to be found in their root systems – and this is a response to where they grow in relation to local drainage and tidal conditions. There are three basic types of mangrove root systems, although some plants are variations from this scheme and have a combination of two root types.

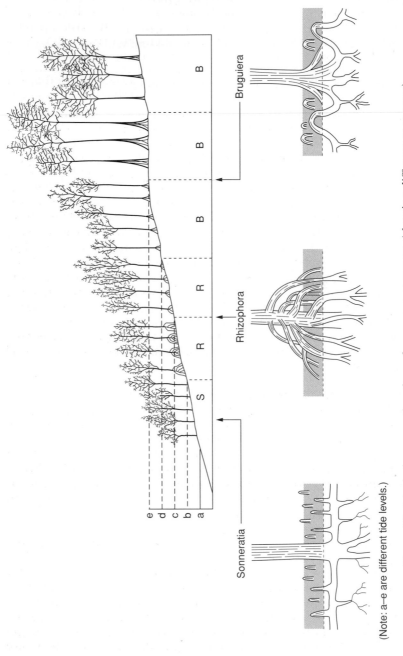

(Note: a–e are different tide levels.)

Figure 13 A typical cross section through a mangrove swamp (showing different root types)

Figure 13 shows a typical cross section through a mangrove swamp from the better-drained higher ground through to the most permanently waterlogged land close to the low-tide mark. The three types of root system can be seen as a sequence through this cross section. Closest to the low-water mark are mangroves with horizontal root systems such as those of the *Sonneratia* genus. Their roots spread out horizontally within the anaerobic mud and send up pencil-like shoots known as *pneumatophores* in order to get oxygen. Even at the lowest tides these pioneering species tend to have their roots partially covered in salt water. Further inland are species that have prop-roots or aerial roots, which give the mangroves the appearance of being on stilts. Once again the purpose of this root type is to give the plants oxygen in an otherwise waterlogged and therefore anaerobic edaphic environment. The mangroves of the *Rhizophora* genus fall into this category. Further upslope and away from the area, which is still waterlogged at low tide, are mangrove species that have root systems that are more like the buttress roots common to many of the larger trees of the tropical rainforest, nevertheless parts of their root systems are adapted to gaining oxygen at times when the tides are high. Mangroves of the *Bruguiera* genus are an example of this type of root system. Both prop and buttress root systems also help to anchor the mangroves, preventing them from being washed away in the outgoing tide. Mangroves are great colonisers and reclaimers of muddy areas and create new land, which can be eventually colonised by other species. Thus, in many parts of the tropics mangroves are the botanical forerunners of much more complex ecosystems such as tropical rainforest.

(ii) The degradation and destruction of mangrove swamps

Although mangroves are hardy plants that can colonise tropical coastal mudflats with comparative ease, various factors are causing their destruction in the contemporary world.

Population pressure and the need for firewood and building materials have led to considerable clearing in rural areas in LEDCs. The introduction of prawn farming either for dietary improvements or for commercial enterprises in many LEDCs in both Africa and Asia has also led to the degradation or destruction of mangrove ecosystems.

Urbanisation, the need for larger port facilities, land reclamation, and the development of tourist resorts and hotels are all taking their toll on mangrove swamps in LEDCs and MEDCs alike. Singapore, for example, was once an island almost completely surrounded by mangroves, but its huge infrastructural, housing, port and tourist developments over the last 40 years have left the mangroves restricted to a few peripheral creeks and to nature reserves such as the Sungei Buloh reserve on the north coast of the island, where boardwalks have been constructed through the swamps.

Pollution is also a great threat to mangroves as their pneumatophore roots can be easily clogged up by oil slicks and other chemical effluents that may float on the water.

4 Rocky Coastlines

Rocky shorelines support lithosere ecosystems that are limited in their biodiversity. They take two basic forms:

(i) cobble or pebble beaches
(ii) bare rock surfaces such as cliffs and flat rocks.

(i) Pebble beaches are one of the least successful environments for plant colonisation, yet a limited number of species are able to survive upon them. The main problems facing colonising plants are the instability of the pebbles themselves, particularly under storm conditions, the lack of fresh water and the lack of light getting to any organisms under the pebbles. Research carried out by Morton (1998) on the wave-exposed cobble shore at Barro Vermelho, on Graciosa Island in the Azores, showed that there are variations in biodiversity between individual berms. The highest landward berm had the greatest variety of plant species as they are able to get a roothold on the more stable cobbles. Species found here include golden rod (*Solidago sempervirens*), rock samphire (*Crithmum maritimum*), Azorean spurge (*Euphorbia azorica*), sea campion (*Silene maritima*) and Azorean bellflower (*Azorina vidalii*). Most of these plants are either halophytes adapted to withstand salt from the seaspray or xerophytes that can cope with a lack of fresh water. Some of these plants such as the rock samphire and Azorean spurge are adapted to cope with both situations. As one moves seawards down the beach, the number of species diminishes until there are no plants growing on the berm closest to the sea.

On shingle ridges in Britain, vegetation is similarly almost absent from the ridges closest to the sea, especially on their seaward slopes. Further inland, however, there are often abundant supplies of fresh water from springs, marshes or lakes and they enable a wide range of plants that are not halophytes to grow between the pebbles where humus can form. Characteristic plants include the yellow horned poppy (*Glaucium flavum*), sea campion and red fescue (*Festuca rubra*). Chesil Beach in Dorset and Dungeness in Kent are two extensive areas with pebble ecosystems.

(ii) Cliffs and flat rocks may offer even more restricted opportunities for plant colonisation than pebbles and cobbles. The vegetation in these rocky areas when out of the reach of the sea spray is open, scattered and surprisingly diverse. Plants establish themselves in cracks and on ledges where some humus can accumulate. Lichens are often abundant as a source of humus formation. Plants found along British rocky shores include sea beet (*Beta maritima*), rock samphire and sea lavender (*Limonium vulgare*). Some rare and localised plants are found in these coastal locations as they are protected by their relative isolation, e.g. sea cabbage (*Brassica oleracea*) and queen stock (*Matthiola incana*). Unlike some of the other types of seres found in Britain, these rocky coastal ones tend to lack any distinctive zoning of plants.

In contrast with Britain, other climatic zones of the world may have much more developed lithoseres. Mediterranean cliffs and flat rocks often have a much greater profusion of plant life, given the range of drought-adapted plants to be found within that climatic zone. Many Mediterranean islands support dense *maquis* or *garrigue* secondary vegetation made up of a wide range of bushes and shrubs. In tropical areas, as mentioned in an earlier chapter, cliff faces frequently support dense vegetation right down to the water's edge.

Certain newly created rocky shores provide 'outdoor laboratories' in which the development of seres can be studied and monitored. Volcanic eruptions, especially when on islands, create new areas of coastline. The gradual colonisation of the ash laid down during the emergence of Anak Krakatoa (son of Krakatoa) in Indonesia following the catastrophic eruption of Krakatoa itself at the end of the 19th century, Surtesey, the Icelandic island that emerged in the 1960s, and the various eruptive stages that have built up Nea Kameni in the Santorini caldera in Greece since the 18th century are all sources of evidence that show how coastal lithoseres develop through time.

5 Coral Coastlines

Coral coastlines are found in around 100 countries and fall into three main **formations**.

- The huge Indo-Pacific formation that stretches from the western coast of the Americas, through the thousands of island groups of the Pacific Ocean, along the northern coasts of Australia (where it includes the Great Barrier Reef), through Indonesia, the Philippines and other parts of South-east Asia, and through the various island groups of the Indian Ocean such as the Maldives and Seychelles to the coasts of Tanzania, Kenya and Somalia in east Africa.
- The Western Atlantic formation that includes the islands of the Caribbean, the north-west coast of South America, the west coast of Central America (where the world's second largest barrier reef is located) and the Florida keys in the USA.
- The much smaller Red Sea formation stretching from the Horn of Africa to the Gulfs of Suez and Aqaba. Despite its size, this formation has some of the clearest waters in the world (due to the region's desert climate) and therefore is often regarded as the world's most favoured area for scuba divers.

a) The development and structure of coral ecosystems

Coral reefs are the largest structures to be created by organic activity to be found on earth; they are also one of the oldest forms of ecosystem, dating back at least 500 million years.

The coral is a small animal, a polyp, that has tiny tentacles for the purpose of feeding from microscopic zooplankton. Each polyp secretes a calcareous skeleton that remains behind when it dies. This is how reefs of coral build up through time and eventually achieve a thickness of hundreds of metres; growth rates are usually between 1 cm and 100 cm per year. Coral skeletons have a great variety of forms such as spherical, cup-shaped, star-shaped fan-shaped and pillar-shaped. The symbiotic relationship that corals have with tiny algae, known as **zooxanthellae**, contributes to the great variety of colours found in the living coral reefs.

The ideal conditions for coral growth are:

• water up to 100 m deep
• clear seas to allow light penetration
• water temperatures averaging 23–29°C
• relatively strong wave activity
• sea salinity of between 30 and 40 ppt.

Corals are only one part of the reef communities as they create a natural environment for a huge variety of algae, reef fish, large predators such as sharks, crustaceans, cephalopods, shellfish, turtles and marine mammals, making them some of the most complex ecosystems on earth.

Reefs develop in three different ways to form:

• fringing reefs, which are those immediately surrounding the shoreline
• barrier reefs, which have a lagoon between them and the mainland
• atolls, which are rings of coral reef surrounding a lagoon.

The theories of how these different formations have evolved are connected with sea-level changes and will be dealt with in the next chapter.

b) The destruction and protection of coral

There are many ways in which coral reefs and other coral formations are under threat, and above all it is through education of people living in tropical areas where coral is found, as well as visitors coming from outside, that will help them to be preserved. The threats come from two main sources: natural processes and human ones.

(i) Natural threats

There are three major natural threats that can lead to the destruction of coral communities: predators, coral bleaching and sea-level rise.

Certain marine animals consume coral, and at the present time the greatest threat is coming from the crown-of-thorns starfish (*Acanthaster planci*). This venomous creature with around 20 spines has occasional population explosions that lead to attacks upon reefs in the Pacific and Indian Oceans, such as the one that has been causing damage to the Great Barrier Reef in Australia from the mid-1990s onwards.

Coral polyps rely upon zooxanthellae green algae, and under certain conditions expel these and lose colour. This bleaching of coral may be the result of several different processes, which include:

- variations in the amount of fresh water running off the land into the sea; more fresh water and therefore lower salinity leads to the loss of zooxanthellae
- increased turbidity and therefore the amount of mud in the water may also follow on from heavy runoff from the land caused by a high incidence of storms
- changes in sea temperature, either above or below that in which coral thrives; such changes are brought about by oscillations in normal ocean current directions.

ii) Human threats

In many coastal coral environments, local people have traditionally taken the coral formations for granted and have used them as a resource base, for both fishing and as a raw material for building. The former is not necessarily destructive as long as over-fishing does not take place. The latter practice, which involves the quarrying of coral for building blocks and as a raw material that is put into lime kilns to produce either limewash or fertiliser, is much more potentially destructive. The Indonesian islands fringed with coral, such as Lombok and Zanzibar island off Tanzania, are both examples of places where population growth is high and the demands for more housing are putting too great a strain on the coral reefs as extraction is taking place much more rapidly than the growing coral is able to replace itself. The increasing, and generally illegal, use of explosives for fishing and coral extraction in many LEDCs is also exacerbating the situation.

Industrial and other forms of pollution is becoming a major problem in the more densely populated areas that have coral seas, such as South-east Asia. Nutrients from sewage, fertiliser runoff and detergents with phosphates in them stimulate the growth of aquatic plants and this leads to the reduction of dissolved oxygen in the water, or the process of **eutrophication**. This leads in turn to increasing turbidity of the water, which is damaging to coral growth. Oil refineries, drilling platforms and oil spills are potentially major sources of marine pollution in coral sea areas of the Caribbean, the Gulf and Arabian Sea, as well as South-east Asia. Oil pollution can lead to the poisoning of reef fish, which has a knock-on effect on coral development.

The growth of tourism and particularly fishing, snorkelling and diving activities in tropical waters is also putting stress upon the coral reefs. The setting up of marine National Parks and other protected areas is one of the most important steps in the move to preserve coral reefs and to extend public awareness. The spread of the ban of spearfishing is one of the main ways in which the coral ecosystems are being protected. Equally destructive if it is carried out by a large number of visitors is the collection of coral or shells, whether dead or alive. Throughout the Red Sea coast of the Sinai Peninsula in Egypt signs state clearly to the public 'Attention. Spearfishing and Coral Picking are not allowed'.

On the islands of the Tunku Abdul Rahman National Park, off the coast of Sabah in Malaysian Borneo, signs discourage shell and coral collection and the causing of pollution through litter by stating 'Leave nothing but your footsteps in the sand; take away nothing but your memories and your photographs'.

Perhaps the most poignant message to visitors is that found in the National Museum of the Marshall Islands, on Majuro Atoll in the Pacific Ocean, where in a glass case containing samples of different shells and corals is a small sign that states, under the heading of 'Note from the Donor': 'To my mind visitors breaking living coral here is like smashing a cathedral window in Europe, it is destroying the irreplaceable'.

Dead coral exposed at low tide, Kuta Beach, Lombok, Indonesia

Summary Diagram

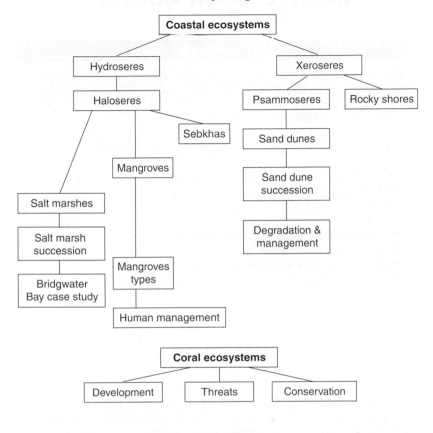

Questions

1. **a)** For either a psammosere or a halosere, draw a labelled diagram that shows the plant sequences which can be seen within it.
    b) Give a detailed account of why this sequence develops.
2. **a)** Compare and contrast the main similarities and differences found in any two coastal ecosystems.
    b) Examine the reasons why different stretches of coastline develop different ecosystems.
3. **a)** Why are many coastal ecosystems so fragile?
    b) Select a coastal ecosystem and analyse its main threats.
4. **a)** In what ways do humans cause degradation of coastal ecosystems?
    b) Explain ways in which human action can help to repair these ecosystems.

5 The Impact of Sea-level Change upon Coasts

KEY WORDS

eustatic change an absolute, rather than relative, rise in sea level
isostatic change change in the relative levels of different portions of the earth's crust
bradyseismic change change in sea level brought about by volcanic or earthquake action
Holocene the period since the last Ice Age (from 11500 years ago onwards)
ria a drowned river valley system resulting from a rise in sea level
atoll a ring of coral islands resulting from sea-level rise

Immense, and coming with crests on fire
We could not understand
Finding the sea so high above the land,
What held its waters from flooding the world entire.

Zennor, Anne Ridler

1 Types of Sea-level Changes

The effects of sea-level change are one of the most complex aspects of coastal geomorphology. Changes are constantly taking place over both the short term and the long term and some parts of the world are more affected than others. Within individual countries too the changes are greater along some stretches of coastline than along others. Sea-level change can be put into two major categories: **relative sea-level rise** and **relative sea-level fall**. The first category includes coastlines where either the sea level rises or the land level falls; the second situation involves either a fall in the level of the sea or a rise in the level of the land. The circumstances in which these changes take place and the causes of them are numerous. Various geographers and geologists have attempted classifications of sea-level change that take into account these different processes.

Bird (2000) gives one of the most comprehensive classifications of sea-level changes, putting their causes into six broad categories:

- eustatic changes
- steric changes
- changes caused by sedimentation
- changes caused by tectonic processes
- geoidal changes
- changes due to human activity.

Each of these is of considerable significance either over the world as a whole or within certain regions, and needs to be examined in more detail.

a) Eustatic changes

Unlike the other categories of sea-level change, eustatic change refers to the absolute, rather than relative, rise in sea level over the very long term. Throughout the thousands of millions of years of geological time, the amount of water on the earth's surface has increased through vulcanism and the development of the gases in the atmosphere. Although there have been short-term falls in sea level due to factors such as the amount of ice cover during the Pleistocene Ice Ages, these were merely fluctuations that buck the overall trend throughout the history of the Earth. These shorter term fluctuations due to ice cover are sometimes known as glacio-eustatic changes. Eustatic change was once thought to be uniform throughout the world, but it is now known to vary regionally.

b) Steric changes

These are the changes which are caused by the changing density and volume of the water contained within the oceans. Rises in atmospheric temperatures warm up the seas causing water to expand and sea levels to rise; the increased salinity of warmer seas also causes expansion. It is estimated that an increase of global temperature of 1°C can lead to a sea-level rise of 2 m. A drop in global temperatures will have the opposite effect and bring about a fall in sea levels.

c) Changes caused by sedimentation

All of the different materials that end up in the oceans as a result denudation of the land – whether from rivers, glaciers, aeolian processes, mass movement or coastal process – will reduce their storage capacity and lead to relative sea-level rises. It is estimated that if all the land above sea level were to be eroded and then deposited into the oceans there would be a global sea-level rise of 250 m; this, of course, is purely hypothetical and could never happen in reality.

d) Changes caused by tectonic processes

A wide range of tectonic processes can have an impact upon relative sea levels and may operate on both regional and local scales. There are four main categories of tectonically induced sea-level changes.

(i) **Epeirogenic changes** are those associated with the changing shape and form of the ocean basins and the land surrounding them. There is evidence that parts of the Pacific Ocean floor are subsiding and therefore the ocean is getting deeper and increasing its capacity. At the same time the process of sea-floor spreading is increasing the capacity of the Atlantic through lateral growth at a rate of a few millimetres per year. There were also upward tectonic movements of certain continents, notably Africa, during Mesozoic times that led to relative sea-level change. These earth movements often took the form of rifts that have given the ocean edges a stepped profile.

(ii) **Orogenic movements**. The processes associated with mountain building, which in turn result from tectonic plate movements, are complex. They involve uplift, tilting and folding of the land both above and below the ocean surface, and are therefore responsible for a variety of sea-level changes. These processes are most active along the newer plate margins or **neotectonic regions** such as around the Pacific rim, and influence areas such as Japan, the Kamchatka Peninsula of Russia and the western coastline of North America. During the Alaskan earthquake of 1964, for example, there was an 11 m uplift of land close to the epicentre, which caused the coastline to advance by over 400 m.

(iii) **Isostatic movements** are changes in the relative level of portions of the earth's crust as a result of loading and offloading of heavy materials placed upon them. The weight of deposits such as deltaic materials and layers of volcanic ash and lava may push areas of crust downwards towards the mantle causing a relative rise in sea level. During the Pleistocene Ice Ages, the sheer weight of the ice cover pressed large areas of crust downwards. With the melting of the ice sheets the land began to recover its former level by rising up again, effectively creating a relative fall in sea level. This process is known as **isostatic recovery**.

(iv) **Volcanic movements**. Areas of vulcanism close to the sea are liable to dramatic coastal changes over short periods of time. The appearance and disappearance of volcanic islands or the gradual increase in size of coastal and island volcanoes are obvious examples of vulcanism creating relative changes in sea level. Underground volcanic activity can also initiate relative sea-level changes, as in the case of the Pozzuoli area near Naples. This phenomenon known as **bradyseismic movement** will be looked at in more detail later on in this chapter.

e) Geoidal changes

These are changes associated with a whole range of phenomena that give the oceans a far from smooth surface. There are troughs and bulges in the ocean surface, which are as great as 90 m below or above the mean global sea level. Gravitational pull of the sun and moon and therefore tidal differences, the underlying topography of the ocean floors, the **Coriolis effect** from the daily rotation of the earth and the circulatory pattern of ocean currents are all believed to be partly responsible for these geoidal changes.

f) Changes due to human activity

Human activity can influence relative sea levels in numerous ways and on a whole range of different scales. At the local level, deliberate actions such as the reclamation of wetlands and the creation of large-scale sea-defences have frequently been carried out in order to create new areas for agriculture or industry. In doing so they have produced areas of land that are actually below sea level, effectively reversing the natural order of things (e.g. in the Fens of England and Polders of the Netherlands). On the other hand, the building of heavy structures upon low-lying coastal areas has led to the depression of land causing a relative rise in sea level (e.g. Venice and the other islands of the Venetian Lagoon). At a global scale, human activity may be having a profound long-term influence upon sea levels through the emission of greenhouse gases. This will be considered in more detail later on in this chapter.

2 Sea-level Changes in Recent Geological Times

The Pleistocene Ice Ages started around 1.8 million years ago, and during its climatic grip over the earth there were 17 major shifts in temperatures with cold **glacials** interspersed with warmer **interglacial** periods. Each glacial period lasted up to 100 000 years, whereas the interglacials were considerably shorter – only around 10 000 years in duration. These changes obviously had a great impact upon the amount of water held in the seas, and throughout the Pleistocene there were alternate rises and falls in sea levels.

Evidence of many of these fluctuations is difficult to find, particularly where sea levels have, on balance, risen and all the landform clues are deep under water. Where, overall, the land level has risen or the sea level fallen, there is more visible geological evidence available.

The last glacial started around 70 000 years ago and ended around 11 000 years ago. Coming out of this last ice age was not a simple progression. Ice cover reached its peak around 18 000 years ago and the climate became gradually warmer, causing ice to melt and sea levels to rise. At around 13 000–11 000 years ago, however, there was another cold period accompanied by ice advance and consequent falling sea levels.

The **Holocene** is the name given to that period since the end of that last mini ice advance; it started around 11 500 years ago and marked the final progression out of the Pleistocene and the influence of large-scale ice cover. A period of rapid sea-level rise came between 11 500 years ago and 6000 years ago, and this is called the **Holocene transgression** or **Flandrian transgression**. As the bulk of the ice caught up in the Pleistocene Ice Ages had melted by 6000 years ago, sea-level rises have been much slower since then. Glaciologists still do not agree about the finer details of rates of sea-level change during the Holocene and many debates continue.

The geographical distribution of the relative sea-level changes is complex, but, on the whole, reflects the proximity of an area to the zones of maximum ice thickness. Thus, places very close to the centre of ice cover such as Greenland, Iceland, parts of northern Scandinavia and north-east Canada are more influenced by a relative rise in the land through isostatic recovery.

By contrast, locations that were a long way from the extended ice sheets such as Australia, South America, much of Africa and southern and south-eastern Asia experienced eustatic sea-level rises and were under no influence of any isostatic changes. It is the areas in between these other two, such as north-western Europe (including the British Isles), the eastern seaboard of the USA, and the northern and eastern seaboards of Siberia, that the pattern is much more complex because they have experienced a mixture of sea-level rise and isostatic recovery. This occurs not just within large regions, but also within smaller localities – for example the South Devon coast has places where emergence has taken place within a few kilometres of features resulting from submergence.

3 Sea-level Rises and Coastlines of Submergence

The greatest amount of coastal submergence that has taken place in recent geological time has been that associated with the melting of the ice sheets, which made major advances during the Pleistocene Ice Ages. As discussed above, the main period of sea-level rise which influenced the world's coastlines in recent geological time was during the first phase of the Holocene transgression from around 11 500 years ago down until about 6000 years ago. One of the biggest effects of this eustatic sea-level rise was the inundation of earlier coastlines and the creation of various types of large inlets and sounds: **rias**, **fjords**, **drowned longitudinal coasts** and **drowned estuaries** (see Figure 14). As well as creating these features, sea-level rise has been held responsible for the development of various coral reef types and, in particular, **coral atolls**.

a) Rias

Ria is the Spanish word for an estuary or river mouth and it has come into English as a technical geographical term because of the many examples of rias on the coast of Galicia, north-west Spain (e.g. Ria de Pontevedra, Ria de Vigo).

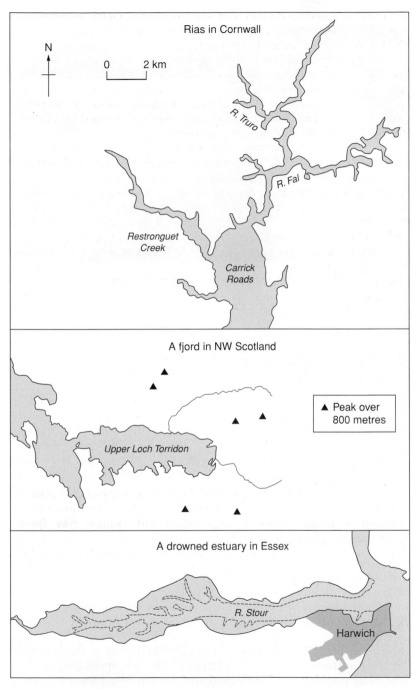

Figure 14 Features of submergent coastlines

The Pontevedra ria, north-west Spain

Rias are formed by the drowning of the lower reaches of a river valley and its tributaries. During the Holocene transgression many coastlines developed ria systems. They tend to occur more along Atlantic type coasts where the geological trend is at right angles to the shoreline and where ridges of hills form the watersheds between the ria systems. In the British Isles rias are most well formed in Pembrokeshire (all the rivers draining into Milford Haven), south-west Ireland (e.g. Bantry Bay), and along the southern coasts of Devon and Cornwall (e.g. the Tamar, Dart, Fal and Fowey estuaries). Those in Wales and Ireland are steeper sided and deeper than those in the south-west peninsula, which reveal vast mudflats at low tide. Rias are also common in Brittany, in France, south-east China and parts of the north-east seaboard of the USA.

b) Drowned longitudinal coasts

Longitudinal or Pacific coasts may also undergo submergence but, instead of forming rias, develop into much broader, open sounds with ridges of land or islands between them. These larger inlets include the Inland Sea of Japan, the area around Vancouver Island in British Columbia, Canada, and stretches of sea between islands in archipelagos such as the Philippines and the Solomon Islands. The most dramatic stretch of drowned longitudinal coast in Europe, however, is the Dalmatian coast of Croatia where numerous long, narrow islands lie parallel to the mainland and are separated from it by deep sounds.

c) Fjords

Fjords (or fiords) are similar to rias but they are the result of the inundation of glacial 'U'-shaped valleys rather than the more common 'V'-shaped river valleys. The word fjord comes from the Norwegian for estuary or river mouth and is related to the Scottish term 'firth'. Fjords can result from either rising sea levels or falling land levels, but in most cases the major fjord coastlines of the world developed at the time of the Holocene transgression.

Fjords characteristically have steep sides and headwalls, flat bottoms containing fairly shallow water and even shallower water where they flow out into the sea. In some cases the seaward 'lip' or 'threshold' of the fjord is exposed and forms islands known as **skerries**.

Fjord coastlines are best developed along high-latitude coastlines where glaciation was and is most effective. Along the west coast of Norway is one the most fjord-indented coastlines in the world; this includes the Sognefjord, which is over 150 km long. Other parts of the world where there are dramatic fjord coastlines include Greenland (e.g. Independence Fjord, Keiser Franz Joseph Fjord), British Columbia in Canada (e.g. Douglas Channel), the west coast of the South Island of New Zealand (e.g. Milford Sound, Sutherland Sound) and the southern coast of Chile (e.g. Seno Europa, Seno Penguin).

In Britain, many of the inlets in north-west Scotland are technically fjords, such as Loch Broom and Loch Nevis. In Denmark, along the east coast of the Jutland Peninsula there are many inlets, which are called fjords (e.g. Hosens Fjord and Vejle Fjord); these are, in fact, rias but the local name for an estuary is used in their names.

d) Drowned estuaries

Low-lying coastal areas that are affected by submergence through rising sea levels without being totally inundated may develop features similar in origin to rias but with a rather different appearance. Drowned estuaries are broad, open estuaries with rather straight and parallel sides. Like rias, they have extensive mudflats cut into by the river channels at low tide, but they lack the dendritic pattern of tributaries that are found in ria systems. There are several good examples of drowned estuaries in Essex (e.g. the River Stour and the River Blackwater). On the other side of the North Sea the estuary of the Ems, which forms the boundary between Holland and Germany, displays similar characteristics.

e) Coral atolls

Coral atolls are remarkable features as they stand isolated within the great areas of tropical ocean, and rise just a few metres above sea level. They are often idyllic places, which fit the stereotype image of a tropical paradise.

Atolls take the form of an almost continuous ring of coral reef around a central lagoon or, more frequently, a series of small islands that form a broken ring around the lagoon. At low tide, the separate islands are often joined by sand-banks. What is strange about atolls' location is their complete isolation from any other major landmass, and their origin was therefore regarded as being rather enigmatic. During his voyage on the *HMS Beagle*, between 1832 and 1836, Charles Darwin made observations about living corals. He realised that they only grew in relatively shallow waters yet they accumulated coral reef material of great thickness.

This led to Darwin coming up with his theory that barrier reefs and atolls both evolved as a result of the subsidence of a volcanic island fringed with coral reefs (therefore a relative sea-level rise). Three formations of corals: **fringing reefs**, **barrier reefs** and **atolls** can be regarded as three stages in atoll development. In the first stage of development, there is a volcanic island within the ocean around which fringing reefs form along the island's edge, in the second stage when sea-level rises cause the island to become partially submerged, the reef continues to grow but does so at a distance of a few hundred metres up to several kilometres offshore in the form of a barrier reef. Over a longer period of time when the island has become completely submerged, but the coral continues to grow, it does so in the form of an atoll: a ring of beaches and reefs located very close to sea level and often broken up into a series of islands or *motus*. Within the Pacific Ocean there are many examples of islands in each of these stages of atoll development:

- fringing reef stage: e.g. Tahiti Island, French Polynesia; Tongatapu Island, Tonga
- barrier reef stage: e.g. Bora Bora, French Polynesia; Chuk Island, Federated States of Micronesia
- atoll stage: e.g. Tarawa Atoll, Kiribati; Majuro Atoll, Marshall Islands.

Darwin's theory still has a lot of acceptance today and work carried out on the Eniwetak atoll in the Marshall Islands in the 1960s tends to back it up. Drilling down into the atoll found there to be an accumulation of old shallow-water coral 1250 m thick, and this was perched on top of a volcano rising over 3 km from the ocean floor.

In the early 20th century the geologist Daly came up with another theory, which was nevertheless also based upon sea-level change. He argued that the basic underlying structure of an atoll was formed during the ice ages when sea levels were low and that once the sea levels started to rise, the coral recolonised the island, growth was reactivated and kept up with the rates of sea-level rise.

Neuman and McIntyre in 1985 identified that the situation was much more complex than either Darwin or Daly had envisaged and argued that coral reefs have one of three strategies in dealing with relative rises in sea level:

- the 'keep-up' reefs, which grow at the same rate as sea-level rise
- the 'catch-up' reefs, which become submerged when sea levels start to rise but are then able to accelerate their growth to cope with the sea-level rise
- the 'give-up' reefs, which fail to cope with sea-level rise and consequently drown.

More recent research and resulting theories have involved much more complex forms of atoll development as a result of the various sea-level fluctuations between glacials and interglacials. However, the basic premise of Darwin still provides a good, simple explanation of reef and atoll growth.

4 Land-level Rises and Coastlines of Emergence

The most significant rises in land levels, which have influenced coastlines in recent geological times, have been those associated with the melting of the ice sheets and the process of isostatic recovery. The current thickness of the world's largest ice sheets over Greenland and Antarctica is around 3 km; during the Pleistocene ice cover would have been considerably thicker over the northern parts of North America and Europe. The great weight of ice caused parts of the crust immediately under the sheets to be depressed down towards the earth's mantle. With the gradual melting of the ice sheets there has been a gradual recovery of the old land levels by the crust rising upwards. This isostatic recovery is still taking place today in many high latitude regions of the Northern Hemisphere. The most notable areas of land-level rise in Europe are around the Gulf of Bothnia, between Sweden and Finland. Here the land is still rising at a rate of up to 8 mm per year. In Britain the highest rates of isostatic recovery are over an area that runs from the central parts of the Scottish Highlands to the northern coast of Ulster and here the land rises by up to 3 mm per year.

Evidence of coastal emergence is also to be found in areas that were not directly covered by thick ice sheets. In these cases it seems likely that their raised beaches and other features were formed during one of the past interglacials when the sea level was much higher.

a) Raised beaches

Raised beaches, sometimes referred to as **fossil beaches**, are areas of sand, shingle or other types of beach deposits that are found high above the current beach level, on top of cliffs or forming other types of platforms way above anywhere that can be reached by the tides or

Raised beaches with cabbages growing on them, south Devon

waves. It is often possible to date raised beaches by the types of deposits they contain; for example, they may have the remains of sea shells of creatures suited to warmer or cooler climates than at present.

The highest raised beaches are those found close to the centre of former ice cover. Along the shores of Hudson Bay in Canada, the highest raised beaches are 315 m above present sea levels; along the Gulf of Bothnia at Skulleberget in Sweden, the beaches reach 286 m above the present sea level. One of the highest raised beaches in Britain is that on the Isle of Mull in Scotland, which is 37 m above sea level.

Most of Britain's raised beaches and particularly those in the south are not from isostatic movements as much as past sea-level rises and falls. Around the Gower Peninsula in South Wales and along the south coast of Devon there are many examples of raised beaches that are 5–10 m above the current sea levels.

b) Other features of emergence

As well as raised beaches, other features may occur as a result of coastal emergence. Shore platforms can be raised up and in many parts of the tropics there is evidence of this where notch and visor type landforms on shore platforms are perched high above the high-tide mark, indicating that the notch is no longer being cut by the sea. Barrier beaches, discussed in the previous chapter, may also develop as a result of a fall in sea level. Certain types of cliff, such as those with a bevelled profile (see Chapter 3), may evolve as a result of a relative rise in the land level.

5 Sea-level Change Through Vulcanism

Coastal areas close to volcanoes and other forms of vulcanism are liable to changing relative sea levels. When dramatic eruptions occur in marine volcanoes as at Krakatoa in Indonesia in 1883 and on Sanotini in Greece in 1470 BC, the whole configuration of land and sea is altered and tidal waves are generated. Marine volcanoes may come and go. In July 1831 a new volcano appeared off the south coast of Sicily, near the port of Sciacca. The British claimed it and called it Graham Island, the French claimed and called it Julia, and the Kingdom of Two Sicilies claimed it and called it Ferdinandea after its king. An international crisis was averted when it exploded and disappeared back into the sea in December 1831.

The presence of magma not far beneath the surface of the earth can also have an impact upon the levels of land and sea: this is known as **bradyseismic** change. The word 'bradyseismic' comes from the Greek and means pertaining to a 'slow tremor'.

CASE STUDY: BRADYSEISMIC CHANGE IN POZZUOLI, ITALY

Pozzuoli is a city of some 80 000 people located in the Campi Flegrei (Phlegraean Fields) volcanic complex to the west of Naples, in Italy. Here there are numerous old craters and some crater lakes, and close to the city centre is a **solfatara** with hot springs, bubbling mud and sulphur-producing fumaroles. Two aspects of vulcanism affect the land and sea levels in and around Pozzuoli:

* the thin crust at this point responds in a rather elastic way to pressure build up in the shallow magma chamber below
* shallow aquifers also respond to the heat and pressure with a build up of steam.

The first major source of evidence of bradyseismic change came in the 1780s when excavations were carried out on the old Roman market place (often miscalled the Temple of Serapis), built around 100 BC. The holes drilled by a stone-boring mollusc, *Lithodomus lithofagus*, were found 5.8 m up on the pillars of the Roman buildings, proving that they had been inundated to that depth at around the 10th century AD. The ruins are now a few metres above sea level. Since the excavations Pozzuoli has received a great deal of scientific attention.

In recent decades there have been two crisis periods brought about by rapid uplift of the land: in 1971–2 the land moved upwards by 1.7 m, then again there was rapid uplift accompanied by over 15 000 minor earth tremors in 1982–4, which caused rises of up to 1.85 m. On both occasions problems occurred because the sewage pipes draining into the sea became tilted upwards towards the land endangering public health. In the 1982–4 crisis 40 000 people were evacuated from the old parts of the city because of the tremors and 8000 buildings were damaged. Since 1984 the land has been subsiding again and Pozzuoli is experiencing a period of calm, although things could change at any time. Figure 15 shows the land level changes during 1982–4.

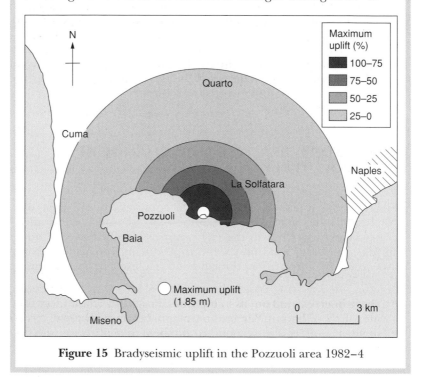

Figure 15 Bradyseismic uplift in the Pozzuoli area 1982–4

6 The Present and Future Patterns of Sea-level Change

During the last two centuries the rates of sea-level rise have been between 1 and 2 mm per year. This is much faster than the average for the last few millennia, which stands at around 0.1–0.2 mm per year. Over this same period of time the global temperatures have increased by around 0.3–0.6°C per century. As this acceleration started in the 19th century it has been linked to industrialisation and the burning of greenhouse gases. There are so many conflicting ideas and statistics about rates of temperature change, but to argue the pros and cons of whether or not long-term global warming as a result of human actions is taking place is beyond the scope of this book.

The international organisation most concerned with getting the right information on global climate change is the IPCC (Inter-governmental Panel on Climate Change), part of the United Nations Organisation. The IPCC uses very sophisticated computer modelling known as **atmosphere–ocean general circulation models** in order to calculate maximum and minimum scenarios of sea-level rise. They also calculate the different contributions made to those changes. At present the IPCC reckons that the minimum scenario annual sea-level rise is at 0.11 mm per year and the maximum rise scenario at 0.77 mm per year. These are predicted to continue for at least 100 years, beyond which it would be difficult to predict. What is striking, how-ever, is the fact that, even on the maximum sea-level rise scenario, thermal expansion of the oceans and seas contributes to about 60% of the sea-level rise, the melting of smaller ice sheets and glaciers con-tributes about 35%, and therefore the melting of the two major ice sheets of Greenland and Antarctica have a minimal influence upon sea-level change. On the minimum sea-level rise scenario, the two major ice sheets do not contribute at all as the amount of snow and ice within the polar regions is estimated to be increasing.

Whatever the causes, the sea levels are rising and there are numer-ous implications for the near future. Sea-level rises are exacerbated in many parts of the world by the fact that as global temperatures rise, climates become more stormy and less predictable. It is therefore not just rising water levels but also the incidence of storm surges that put coastal communities in danger.

In MEDCs there are many concerns connected with flood preven-tion along the low-lying coasts. In some cases large-scale engineering works have been carried out as in the Delta Project in Holland and the building of the Thames Barrier to protect London. Coastal resorts have had to make decisions based on cost–benefit analysis on how best to maintain their seafronts and sea-defences. In Britain yet another concern is the effect of sea-level rises upon nuclear power stations. Rising sea levels and coastal retreat could cause damage to several of the coastal nuclear powers stations in the next few decades, including those at Hinkley Point in Somerset, at Dungeness in Kent, at Sizewell in Essex and the reprocessing plant at Sellafield in Cumbria.

In LEDCs the concerns are greater; not only is there far less money to be spent upon coastal defences, but there are densely populated agricultural lowlands in coastal areas, where rising sea levels threaten the very livelihoods of the people living there. Bangladesh is a country where millions live close to the shore of the Bay of Bengal, upon muddy islands within the Ganges–Brahmaputra delta and along tidal creeks, and would be particularly badly affected by sea-level rises in the future. Some of the coral island nations, particularly where most are made up from atolls, e.g. the Maldives in the Indian Ocean and the Marshall Islands, Kiribati and Tuvalu in the Pacific, are under even greater threat as they have little land that is more than a few metres above sea level.

CASE STUDY: THE IMPLICATIONS OF SEA-LEVEL RISE ON PACIFIC ISLANDS

In October 2001 Andrew Simms wrote an article in *The Guardian* entitled 'Farewell Tuvalu' in which he rather alarmingly wrote 'A group of nine islands, home to 11 000 people, is the first nation to pay the ultimate price for global warming'. Even the *Lonely Planet* website advises visitors to get there soon before the rising sea levels swamp the islands.

Alarmed by the publicity about rising sea levels, in 2001 the Tuvalu government appealed to Australia and New Zealand, their Pacific neighbour 'superpowers', for help and asked for the entire population to be evacuated and be granted environmental refugee status. Australia, with its strict immigration laws, refused the request. The islands' politicians now blame Australia for both not taking their plight seriously and for being a contributor to the greenhouse gases in the atmosphere, which is the claimed source of global warming and sea-level rise.

The main problem is that the sea levels are not actually rising in Tuvalu. The Australian National Tidal Facility has been taking measurements on Tuvalu for several decades and has found absolutely no sea-level rise to have taken place in the last 10 years. What has effectively concerned the inhabitants of Tuvalu has been the increase in beach erosion, but this has been caused by storm surges, which may be on the increase as a result of higher world temperatures. Lying between 3°N of the Equator and 9°S, Tuvalu is actually outside of the normal tropical storm belts.

A similar situation exists in Kiribati, another widely scattered coral island nation made up mainly of atolls, and home to a population of 77 000 people. The highest points on the main atoll, Tarawa, are its bridges, which link the various small islands. These are between 3 and 6 m above sea level, whereas most people live and farm at around 1–2 m above sea level. Traditionally the people have built simple sea-wall defences where it was felt to be necessary and storm surges occasionally cause damage. The building of bridges has to some extent altered the natural balance of tides and currents between sea and lagoon. The most ambitious new bridge is that built between Betio, the busy port island, and its neighbour, Bairiki, the island that is the main administrative centre. The 300-m long Japanese-built bridge has caused some areas of beach to be eroded and others to experience accretion through longshore drift. It has also made some extra storm flood defences necessary for people whose properties have been affected by these coastal changes.

Summary Diagram

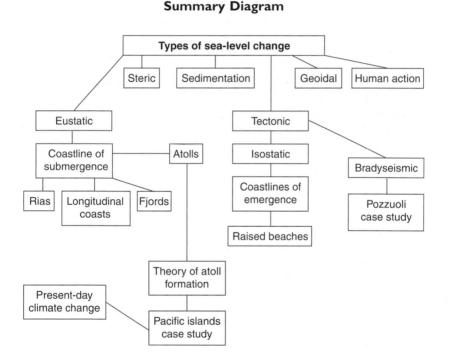

Questions

1. **a)** Attempt a classification of the different types of sea-level change.
   b) Explain the conditions under which these changes occur.
2. **a)** Explain the differences between emergent and submergent coastlines.
   b) Choose two features of submerged coastlines and two of emergent coastlines and, with the aid of diagrams, explain their formation.
3. **a)** Explain the differences between isostatic change and eustatic change in relation to sea levels.
   b) Describe the various features that result from these processes.
4. **a)** Outline the main causes of contemporary sea-level rise.
   b) What are likely to be the main problems associated with sea-level rise in the future?
5. **a)** What do past changes in sea level tell us about the long-term relationships between land and sea?
   b) How might an understanding of this influence human activity in the future?

6 Human Management of the Coastline

> You gentlemen of England
> Who live at home in ease,
> How little do you think
> Of the dangers of the seas.
>
> *The Valiant Sailors*, Martin Parker

The relationships between humans and coasts go back millennia. In the ancient Mediterranean, the Egyptian, Phoenician, Carthaginian, Minoan, Greek and Roman civilisations all had a close relationship with the sea – in the last case so much so that the Mediterranean was referred to as *Mare Nostrum* (Our Sea). Elsewhere in the world, other civilisations such as those of India and China also had strong links with the sea. Large proportions of many countries' populations live in coastal regions and many capital or principal cities have maritime locations. The building of ports, harbours, breakwaters and lighthouses have a very long history, and therefore so too does the human modification of parts of the coastline. The piecemeal reclamation of coastal marshes goes back to Greek and Roman times, and this in many parts of the Mediterranean changed the configuration of the coastline. In more recent centuries reclamation has taken place on a much larger scale and has completely altered the coastline of the Netherlands.

1 Approaches to Coastal Management

Throughout the world, and in particular in MEDCs where the economic and technological resources are available and where more academic studies of coastal processes have been carried out, four main approaches towards coastal management can be identified:

- **letting nature have its way** and letting erosion, longshore drift and other processes take place unchecked
- **'soft'-engineering schemes**, which use natural processes and natural materials to slow down or reverse some of the changes that take place along the shoreline
- **'hard'-engineering schemes**, which use imported materials to change the form of the coastline and make it less susceptible to erosion and other natural processes
- **managed retreat**, which accepts that natural processes will always be taking place and involves moving backwards the limits of the interface between the shore and built environment.

a) Letting nature have its way

From the late 20th century onwards several factors have been mitigating against the use of hard-engineering schemes along coastlines, especially in MEDCs where they had often become the norm. Above all else it has been the escalating costs of coastal defences that has encouraged authorities to re-assess their attitudes and approaches towards the protection of land and property against the action of the sea. In addition to this, the increased environmental awareness of both individuals and society as a whole, the actions and campaigns of local environmental groups and the retreat away from intensive agriculture have led to the re-assessment of coastal defence policies. A further factor that appears to be having an influence on these decisions is the increased occurrence of more turbulent weather, which is leading to the increased incidence of storm surges worldwide.

In Britain, where both coastal and wider flood defences are the overall responsibility of the Environment Agency a Non-Departmental Public Body funded through DEFRA (the Department for the Environment, Food and Rural Affairs), the political tide has turned since the late 1990s when a series of studies and reports came out in favour of abandoning engineering schemes and either leaving the coastline to the processes of nature or adopting a policy of managed retreat. These changes are particularly pertinent to the situation in many parts of southern and eastern England where rocks are softer and more readily eroded, and therefore natural rates of coastal retreat are most rapid. These parts of the country with large areas of flat land close to the sea are also more liable to coastal flooding. Not all interest groups see eye to eye with the government and the Environment Agency. The National Farmers' Union continues to favour the hard-engineering schemes that prevent their members' land from being flooded. At the same time, many local authorities are critical of the changes in policy as they rightly fear that not enough money is available from central government to compensate for any losses of buildings and public works such as roads. At the regional level there are 18 'Coastal Groups', which are advisory bodies made up from various local authorities.

These were set up from the late 1980s onwards partly to replace the old, muddled system of coastal management; some of these groups are more conservationist and environmentally conscious, and others are more in favour of the continued use of 'hard'-engineering schemes. In the long term, however, general acceptance of this approach will have to be made on the grounds of cost as well as environmental concern. In addition to this, there has been a general move towards more emphasis upon the **sustainability** of coastal protection schemes, i.e. they should be successful over a long period of time without too much damage to the environment. Increased emphasis is also being put upon authorities being involved with **Integrated Coastal Management** (ICM), which considers not just the coastal environment but other related local natural environments such as wetlands and watersheds as well as integrating them with policies relating to local economic activities including agriculture, tourism, water resources, transport infrastructure and industry

b) 'Soft'-engineering schemes

Soft-engineering schemes are those that work with nature and natural processes, and tend to blend in better with the natural environment than the hard-engineering schemes that will be dealt with later. These schemes are, needless to say, more favoured by environmentalists and conservationists than the various products of hard-engineering works.

- **Beach replenishment or beach nourishment**. The use of groynes (a hard-engineering process mentioned below) contains beach materials and therefore, to some extent, helps to redistribute them. When longshore drift, acting over a long period of time, has carried away the majority of the material upon a beach, it may be necessary to bring in sand or shingle from where it has been redeposited. In cases where the materials have been lost altogether by being deposited out to sea, sand or shingle may have to be brought in from further afield.

  Beach replenishment is simply carried out by bringing in beach materials by lorry and dumping them upon the eroded beach. This process is particularly important in resorts that rely on their good sandy beaches for tourism. Although expensive to carry out, the effects of replenishment can last for years or decades. The entire beach at Hythe in Kent was replaced in the early 1990s following very extensive erosion. Replenishment was also part of the integrated programme of sea-defence renewal carried out at

Eastbourne, Sussex in the late 1990s. Both resorts have east–west trending beaches on the south coast of England, making them particularly vulnerable to longshore drift created by the south-westerly prevailing winds.

Some examples of beach replenishment are carried out for 'cosmetic' reasons rather than as a remedy for beach erosion. On Tenerife in the Canary Islands there is little naturally occurring 'golden' sand, but the dominant beach material is black sand from the broken down basalt of which the volcanic island is made. In order to enhance popular resorts, such as Los Cristianos, hundreds of thousands of tonnes of paler coloured sands have been imported from mainland Spain and from the nearby Moroccan coast.

One of the potential problems of beach replenishment is the carrying away of material from a place, which then upsets the sediment cell balance in this source area. This was witnessed in the late 19th century at Hallsands in South Devon where offshore dredging of gravel and pebbles for use in the construction of the Royal Dockyard at Devonport led to the increased erosion of the shoreline and the disappearance of most of the coastal village. Today just a few ruined houses are all that remains of Hallsands. One solution to this problem is to take sand or other materials for beach nourishment from sites inland rather than coastal ones.

- **Sand dune construction**. Lines of sand dunes along the coast have long been recognised as excellent barriers to flooding of low-lying land or the erosion of cliffs lying further inland. In the Netherlands not only have coastal dunes been stabilised and enhanced, but they have also been artificially created. Fences of brushwood and wooden pales driven into the ground where loose sand is being blown around act as barriers against which sand piles up and eventually forms into dunes. The optimum conditions for this have been achieved where the fences have a porosity (i.e. the amount of sand they let pass through, vis-à-vis that which accumulates) of 50%. It has also been found that the process works better when there are natural dunes nearby or where a psammosere sequence has already begun (i.e. there are already embryo dunes *in situ*).

- **Sand dune stabilisation and management**. Another way in which soft engineering can be applied to coastal areas is in the management of sand dunes. Dunes shifting inland or along the shore may pose a threat to buildings, infrastructure and farmland. Sand dune stabilisation schemes can be in close harmony with nature by using natural materials. Stabilisation of dunes is important both when they are highly mobile and when blow outs occur. There are two main ways in which stabilisation can be achieved: the first is to fix the dune ridge by putting up fences made of natural materials such as brushwood or wickerwork along it, the other is to plant the dune with colonisers such as marram grass, which are likely to take root and quickly spread over the dune. Also important in dune management in areas that are popular with tourists is to restrict access to the public in order to reduce the erosion of vegetation and the likelihood of blow outs occurring. After decades of damage caused by trampling, the Comune di Roma (the town council of Rome, Italy) has imposed a very strict access policy to the sand dunes parallel to the 5 km stretch of public beaches south of Castel Fusano on the coast of Lazio, close to Rome. Since the late 1990s access from the main road to the beach has been limited to new, regularly positioned boardwalks through the dunes.

c) 'Hard'-engineering schemes

From the 18th century onwards the approach towards coastal management has been dominated by the application of engineering schemes, which have used the latest technology and materials to protect property, ports and harbours, as well as tourist facilities, from being eroded or otherwise altered by the action of the sea. This bolder approach to solving what was perceived as the 'problems' of coastal change coincided with both the Industrial Age and the development of the large popular seaside resorts. These schemes often involve the building of large-scale structures, many of which are both unsightly and fail to blend in with the natural environment. These are therefore called 'hard'-engineering solutions. In some countries a substantial proportion of their coastlines have taken on an artificial character as a result of these 'hard'-engineering works where they are highly developed. Bird (2000) estimates that 85% of the coastline of Belgium, 51% of that of Japan, 38% of that of England (note: not Britain as a whole) and 21% of that of South Korea can be regarded as artificial as a result of a combination of port and seaside resort infrastructure combined with 'hard'-engineering schemes to protect the land from the sea.

d) Types of coastal defences associated with 'hard'-engineering schemes

A wide range of hard-engineering schemes have been put in place in different parts of the world to protect coastlines from cliff or beach erosion, longshore drift, high tides and other natural processes that may disrupt the existing natural or built environment. These various methods of protection may be used individually, but increasingly there is a trend towards using more than one protection scheme in conjunction in order to achieve more lasting success. Figure 16 illustrates many of these types of sea-defences.

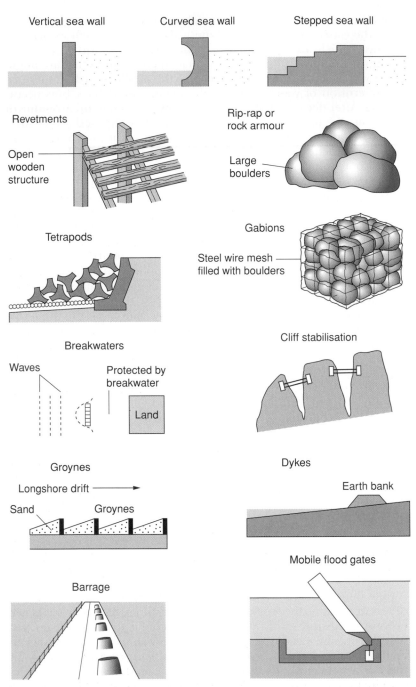

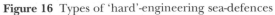

Figure 16 Types of 'hard'-engineering sea-defences

- **Vertical sea-walls**. These have been one of the most traditional ways of protecting harbour sides, promenades and roads in coastal towns and villages. In Europe these types of protective walls go back to at least Greek and Roman times. Typically, modern sea-walls are made of concrete or masonry (the latter is more expensive but more durable), and have wide and deep foundations. The high capital costs of these walls generally means that they are restricted to built-up areas. Although they are effective in holding back most storms, they may be breached and damaged in storm surges. Their biggest design fault is that they can be weakened by both hydraulic action and corrasion as they take the full impact of breaking waves.
- **Curved sea-walls**. Increasingly these are replacing vertical sea-walls in places that can afford to construct them; in the last decade new curved walls have been built at seaside resorts such as Eastbourne in Sussex and Burnham-on-Sea in Somerset. Although more expensive to construct than vertical walls (about double the cost) they are much more durable and therefore annual maintenance costs are much lower. This type of wall is shaped in the same way as a breaking wave and therefore reflects and dissipates waves and so their energy is much reduced. The biggest problem faced by curved walls is that they can be undermined by the scouring effects of waves, which can weaken their base, and they also tend to encourage erosion of beach material immediately beneath them.
- **Stepped walls**. These are yet another variation on the sea-wall. They have both the effect of reflecting and dissipating waves as with the curved sea-wall. The disadvantages of this type of wall are the amount of space needed for it – but this is not necessarily a problem in places with a high tidal range – and the fact that it is even more expensive than curved sea-walls.
- **Revetments**. These are sloping aprons, which encase sections of beach or low-angle cliffs. They are made of various materials from concrete slabs to timber frames. Generally, they are made with rough surfaces or have holes in them in order to dissipate the energy from breaking waves. Although cheaper to build than sea-walls, they generally need to cover a much larger surface area and are very unsightly. Revetments can take a heavy lashing from the waves and, although they may be effective in protecting the shoreline, they themselves may become eroded and need frequent replacement, e.g. the concrete revetments in Mount's Bay near Penzance in Cornwall. In some places on the north Norfolk coast such as Overstrand, revetments have been constructed from tropical hardwoods, which are likely to need replacing after about 10 years; this is very wasteful of important natural resources.
- **Rip rap** or **rock armour**. This is the use of large boulders piled up in front of places that are to be protected, such as cliffs, beaches

and harbour walls. These rock piles are permeable and when waves break upon them a lot of the energy is immediately dissipated. Rip rap is much cheaper than installing sea-walls, especially if a local rock is suitable for its use. On many occasions the places that need protecting are geologically young and the best rocks for rip rap are resistant ones such as granite and basalt, which have to be brought in from elsewhere adding greatly to the costs. It could be argued that rip rap is unsightly but it all depends on how it blends in with the seafront. At Horta, the main urban centre of Faial Island in the Azores, local basalt is used for rip rap along the town's main coastal promenade and this blends in well as it is also used as a building material. By contrast, at many places in Britain harder rocks may be brought in from many hundreds of kilometres away when there are no local rocks of sufficient resistance, such as at Reculver in Kent.

- **Tetrapods** or **dolos**. These 'four-legged' cast pieces of concrete have much the same function as rip rap. Tetrapods have been designed to interlock and stack up into piles so that they can be easily lifted into place by cranes and made into sea-walls or sea-wall reinforcements. They have the advantage over rip rap of being more stable and therefore can be stacked much higher. Tetrapods are also cheaper to buy and cheaper to install than rip rap. The biggest problem about tetrapods is that they are so ugly and are therefore generally a complete eyesore in any picturesque location. Tetrapods are widely used in LEDCs because of they are a very economic form of coastal defence. They are also widely found in former Communist countries of eastern Europe where there were few reservations about the aesthetics of engineering schemes. For example, tetrapods are the main form of sea-defence around the commercial harbour of Costanţa in Romania and along the neighbouring Black Sea beaches. Tetrapods are also used to form the huge wall that separates the Corland spit from the lagoon channel entrance to the port of Klaipéda in Lithuania; in this case it forms a barrier to stop the northward movement of sand through longshore drift.

- **Gabions**. Gabions are wire cages filled with stones, which may vary in size from boulders down to pebbles. Like tetrapods, but unlike rip rap, they are made to a standard size. This means that they can be easily put together to form sea- or harbour walls or to reinforce existing structures. Like tetrapods, they can be an intrusive eyesore. Gabions can be seeded with vegetation, however, to make them blend in better with the natural environment. Gabions can be made into vertical walls, stepped walls or may be used as extra defences for pre-existing masonry walls.

- **Breakwaters**. These walls actually built in the sea are one of the most effective forms of coastal protection, but creating them may involve some complex engineering as areas of sea may have to be cut off and the water pumped out during construction. Breakwaters are a fundamental element in the construction of harbours as they provide the necessary protection to moored vessels. They may be constructed of any of the same materials as sea-walls, such as concrete, rip rap and tetrapods; generally more than one of these types of materials are used in combination to provide thicker walls. For example, the main harbour breakwaters at Angra do Heroismo on the island of Terceira in the Azores, which were reconstructed in 2003, are made of thick concrete walls toped with tetrapods on the outer, seaward side. In the more picturesque fishing villages of the West Country, often only local stones are allowed to be used in order for the breakwaters to blend in with the natural environment, e.g. the harbours at Mullion Cove and Boscastle in Cornwall, at Lynmouth in Devon and at Porlock Weir in Somerset. In beach areas, particularly where longshore drift is taking place, a series of small offshore breakwaters parallel to the sea may be constructed of materials such as rip rap or tetrapods. These structures have the effect of containing the beach materials that build up into a series of cusp-shaped beaches. This method has been effectively used in such diverse locations as the Baltic coast to the south of Arhus in Denmark, the beaches of the Black Sea to the north of Costanta, Romania and along the southern coast of Singapore island. Another effective use of breakwaters is to create offshore reefs to dissipate wave energy hundreds of metres offshore. This can be done by dumping large boulders and so creating rip rap walls in the shallower seas where breaking waves could be a threat to human activities. This form of management imitates what naturally occurs in such places as tropical coral shorelines.

- **Cliff stabilisation**. Whereas the hard schemes listed above act as protection from marine erosion of cliffs, beaches and human structures such as roads, hotels and houses, as well as creating safe harbours, engineering methods are employed for containing cliff profiles and preventing them from further slope failure. These schemes are not unique to the seaside as they are also to be found in mountain areas and along major roads where deep cuttings have been made. Cliff stabilisation is often expensive and can be very temporary as the materials get attacked by subaerial erosion, especially in places where clays, mudstones and shales form the cliffs. Schemes to stabilise cliffs include wire netting or plastic sheeting being spread over the unstable slopes, the forced seeding of the slopes to encourage the growth of vegetation and, in the case of more resistant jointed rocks that are prone to toppling, the securing of the slopes with stays and bolts. Most of these schemes are unsightly and tend to be used only in places where property is at risk. The variety of cheap forms of sheeting and pegging that are used on the soft cliffs above sections of the beach at Santa Monica, in Los Angeles, USA are an example of a most unsightly piece of human interference along a coastal environment.

- **Groynes**. The main function of groynes is to slow down the rate of longshore drift. Groynes are a common sight in seaside resorts and take the form of a series of fences at right angles to beaches and are usually located at intervals of a few hundred metres. Groynes interrupt the process of longshore drift as sand and pebbles pile up against these retaining walls and cannot move further along the beach. Instead of materials moving from one end of the beach to the other, they merely accumulate at each groyne and it is easy for it to be manually or mechanically redistributed. Groynes, the practice of construction of which dates back to the 19th century, were traditionally made of wood. From the second half of the 20th century both wood and concrete have been used. Both materials have a limited life and become weathered and disintegrate. Although concrete is potentially cheaper, wood is regarded as being better aesthetically and many resorts have reverted back to the use of wood in preference to concrete, as is the case at Eastbourne in Sussex

- **Dykes**. Dykes are embankments that protect low-lying coastal lands from the incursion of the sea during storm surges, spring tides or the onslaught of a succession of winter storms. There is a long history of dyke construction in certain parts of Europe going back to pre-Roman times, e.g. in Holland and the Somerset Levels. In periods of population pressure upon the land and when advances were made in drainage technology, large-scale reclamation schemes took place in marshy areas of Europe, creating thousands of hectares of new farmland close to or below sea level. With each new scheme, larger and stronger dykes were required. The biggest changes in both the Netherlands and England came from the mid-18th century onwards, and now vast areas of land are protected by earthen embankments, strengthened in places by other materials, if necessary.

- **Barrages**. Barrages are more solid, hard-engineering versions of dykes and embankments. They can be regarded as 20th and 21st century remedies to coastal flooding. The great coastal floods on the shores of the North Sea that affected both eastern England and the Netherlands in 1953 led to one of Holland's most ambitious hydraulic-engineering schemes – the Delta Project. This sealed off four of the distributary mouths of the Scheldt and Rhine by putting barrages across between islands within the estuarine delta. This has had the multi-purpose effects of creating new farmland and large reservoirs of freshwater, and the creation of new infrastructure, particularly road links, as well as providing coastal protection from future storm surge flooding.

- **Mobile floodgates**. Sluices have been used for centuries to control water flows, especially in areas that have been reclaimed from marshland. These vertical gates, which can be opened and closed, are normally found in conjunction with drainage canals, pumping stations

Small breakwaters creating cusped beaches, Baltic Sea, Denmark

Breakwaters at Mullion Cove, Cornwall

and coastal dykes. As part of the British response to the flooding of the North Sea coastlands in 1953, an assessment was carried out as to having floodgates on the Thames Estuary to prevent central London from flooding in the case of future storm surges. The result of this was the building of the Thames Barrier, which went into operation in 1984. This acts in the same manner as smaller sluice gates but in a more sophisticated way, as a series of floodgates, which are housed under water, rotate into action when necessary and seal the Thames upstream off from the encroaching tides. In the near future, the Venetian Lagoon is to have four mobile floodgates of a different design put into place to prevent Venice and the other settlements of the lagoon from sea surge flooding; this is dealt with in more detail in the next chapter.

CASE STUDY: CHOOSING WHAT ACTION TO TAKE ON THE ISLE OF SHEPPEY, KENT

Cost–benefit analysis has become the most crucial part of the decision-making process about how to extend, replace or abandon coastal defences. The huge and rising costs of sea-defences in the late 20th and early 21st centuries have led to the retreat away from the indiscriminate use of hard-engineering schemes everywhere that needs protection. In Britain, the Environment Agency and local authorities now favour more limited use of expensive schemes and favour managed retreat or total abandonment of sea-defences where appropriate. The Isle of Sheppey in Kent provides a good example of where the coastal management plans have become discriminatory on the basis of cost.

The island is generally low-lying, roughly 15 km long and 7 km wide, and located off the north coast of Kent. The southern side of the island, which is marshy and largely uninhabited, is where Sheppey is separated from the rest of Kent by a muddy creek known as the Swale. The northern coast of the island has a series of settlements along it where there are permanent dwellings, holiday homes and caravan sites. This coast is formed of soft, easily eroded cliffs made of London Clay that reach 73 m at their highest point. The cliffs are being constantly attacked by weathering from rainwater and cracking, from rotational slumping when rain penetrates cracks and from toe erosion during periods with stormy seas. Some groynes and containing concrete defences were historically used along this coast, but following heavy storms in the late 1970s, when coastal retreat caused considerable loss of caravans into the sea at Boarers Run, the whole situation was reassessed.

The local authority and borough engineer brought out a report that considered the fate of six problem areas along the north coast of Sheppey. Of the six locations only two were selected as places where further money would be invested in coastal defence works, and £2.25 million were invested in the replacement of groynes and the construction of a concrete toe wall to prevent the cliffs being further eroded. These two locations, the Leas at Minster and Warden Bay, are the most built up with permanent residences and, in the latter case, a holiday camp. By contrast, the other four locations: Oak Lane, Boarers Run, Eastchurch Cliffs and Warden's Point, were felt to be not worth investing further money on coastal defences, leaving their cliffs exposed to the processes of weathering and erosion. In places such as Boarers Run and Warden Point the cost of new sea-defences would have been prohibitive and it proved much cheaper to move their caravan sites further inland, a form of managed retreat.

CASE STUDIES: THE RELOCATION OF BUILDINGS – THE HARDEST OF HARD-ENGINEERING SCHEMES?

In the spring of 1999 an event on the East Sussex coast attracted a great deal of media attention in Britain, including a special edition of BBC TV's *Tomorrow's World*. The Belle Tout lighthouse, which had been constructed not far from Beachy Head, near Eastbourne, was moved 17 m inland at a cost of around £3 million. When the lighthouse was built in 1834 it had been 35 m from the cliff edge, but 164 years of cliff recession left it 16 m from the edge. Then in 1998 a huge rock fall caused a loss of 13 m of cliffs leaving the Belle Tout precariously positioned 3 m from the edge of a 100-m drop into the English Channel. The private owners of the lighthouse decided to move it rather than allow nature to remove it. The operation involved the underpinning of the lighthouse, the insertion of beams under it and then the placing of 22 hydraulic jacks underneath the building, which enabled it to be carried inland to firm concrete foundations. Considering the rate of cliff recession, the Belle Tout will probably need to be relocated again in about 2040.

This is an example of where cost–benefit analysis seems to be ignored in favour of the conservation of a building; the property value of the Belle Tout would not under normal circumstances justify the high cost of the hard-engineering works involved in its removal. McGlashan (2003), who did a detailed study of this type of **managed relocation**, shows that the Belle Tout experience is far from unique. In his work he considered the Beachy Head example alongside three others, one in Scotland and two in the USA.

(i) The Buddon Lighthouse on the Tay estuary was moved in 1884 because of sand encroachment upon its site. The movement of the lighthouse to a new site 63 m away took 1 month and involved a huge iron strap being placed around the building, and large amounts of timber, lubricants and manual labour. The cost was just £285 (the equivalent of 100 times the cost today) most of which was for the workmen's wages. The lighthouse is still in place today but is no longer used.

(ii) The Brighton Beach Hotel on Coney Island, New York was a huge rambling building in 19th century Gothic style, located on a low-lying coast dominated by sand dunes and salt marshes. Changes in the coastline due to erosion necessitated the moving of the hotel in 1888. The movement was carried out by the laying of rail tracks, the underpinning of the building upon vast timber sleepers and the pulling of hotel inland by steam locomotives. The move cost an enormous US$80 000 and left the hotel in a safe location 151 m from the sea. The Brighton Beach Hotel was eventually demolished in the 1920s, but buildings now on its site are under erosional threat, which is being counteracted by measures such as groynes and beach nourishment.

(iii) In comparison with the three large-scale and historic events described above, there are more commonplace events that take place along the coast of North Carolina. Since the 1960s building regulations have made it necessary for houses built close to the sea to be constructed in a way that makes them easily moveable, there are 30-year and 60-year 'setback lines' behind which buildings can be moved; these are identified on the basis of average coastal recession rates. The average cost of moving a threatened property is between just US$5000 and US$11 000. In terms of cost–benefit analysis, this is extremely cheap as insurance claims for houses damaged by erosion or coastal flooding have averaged US$62 000 in recent years.

2 Multipurpose Coastal Management

Given the wide range of human uses of the coastline, many of the modern forms of coastal intervention have become multipurpose. Certainly, where new coastal defences have been created, several different types of hard-engineering schemes are used in conjunction. Some opportunities exist to carry this forward on a much larger scale. Within estuaries, for example, tides can be harnessed for electricity generation and barrages can be used for this purpose, along with the creation of new roads and the storage of fresh water. In Britain the idea of putting a barrage across the Severn Estuary has long been mooted, yet it has still not been built. If the scheme were to come to fruition, it would probably be the first of many as similar proposals have been made to dam such diverse stretches of tidal waters as the River Mersey, the Wash and Morecambe Bay.

Two factors are at present mitigating against such large-scale engineering schemes and may lead to the Severn Barrage never being built:

- the reluctance of governments and the private sector to invest in such schemes, which may not produce profits for decades (c.f. the Channel Tunnel project)
- the general acceptance in the scientific and environmentalist communities that large-scale multipurpose schemes may in the long term create more problems than they solve (e.g. the Aswan Dam and other 'megadams' such as the Narmada Dam in India and the Three Gorges Dam in China).

CASE STUDY: THE SEVERN BARRAGE PROPOSAL

The power of the tides and the generation of electricity from them have been considered in Chapter 1. One of the most ambitious engineering schemes ever proposed in Britain is that to put a tidal barrage across the Severn Estuary. In both 1849 and in 1933 there were proposals to put a barrage across the Severn at roughly the point where the second road crossing goes today. The first of these projects was purely for a transport route as it predates hydro-electric technology, the second scheme was however a multipurpose one which included power generation.

Although these projects had already been put forward, the most serious and detailed proposal was made in 1974. This was to put a barrage across the estuary some 20 km further downstream where the stretch of water was much wider. The route followed by the proposed barrage was to join the Brean Down headland on the English side to Laverack Point on the Welsh side but incorporating the two mid-estuary islands of Flat Holm and Steep Holm (see Figure 17). Over a 13-year period the proposal absorbed some £67 million in feasibility studies and local public enquiries. The project was eventually dropped on environmental grounds, although it could be revived at any time in the future.

The most recent proposal concerning the Severn Estuary was made in 2003. The British government and the Welsh Development Agency are jointly funding a project, which is setting up a £ 1.5 million prototype tidal generator in the much smaller Milford Haven inlet in Pembrokeshire. If this is successful, it could lead to a whole series of much bigger generators being located in the Severn Estuary and similar places along the west coasts of the UK. Unlike the earlier projects, this scheme, which is being carried by a company called Tidal Hydraulic

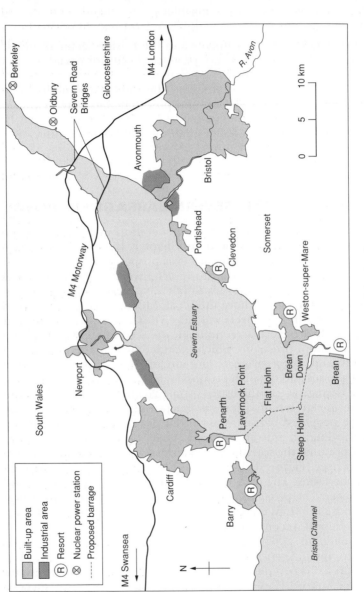

Figure 17 The proposed Severn barrage

Generators, is less likely to upset environmentalists. There would be no dam or barrage across the estuary, but instead the generation plant would sit at the bottom of the seabed. The Milford Haven pilot scheme will only produce 1 MW of electricity, but it is hoped that underwater power stations with a capacity of 50 MW (enough to power 25 000 homes) will soon follow.

This prototype is to be 10 m high and 80 m wide, and will sit in 40 m of water and consists of a battery of propellers, which will turn in the tides; electricity will be generated for 20 hours of the day. The main tidal stream generator if built within the Severn Estuary itself would be much larger and could be up to 17 km wide and sit on the seafloor at a depth of 60 m. Unlike the old barrage scheme, there could be more than one of the propeller-based power stations within the Severn Estuary and Bristol Channel.

The biggest problems associated with any development of the Severn Estuary are the numerous different uses to which the waters and the land along the estuary are currently put and the wide range of interest groups, all of whom have contrasting attitudes and opinions on the building of a barrage.

The Severn Estuary has a great variety of human activities within it and along its shores. There are several large cities, including Bristol on the English side and Newport and Cardiff on its Welsh shores. Heavy industries using large quantities of water are concentrated at Avonmouth near Bristol and between Newport and Cardiff. There are nuclear power stations at Berkeley and Oldbury, between Bristol and Gloucester, and at Hinkley Point in Bridgwater Bay, as well as conventional thermal power stations on the Welsh coast. Barrage construction could well make these older power stations redundant. The barrage would create great reservoirs of fresh water, which would generally be of more use to industry than salt water or brackish water (the availability of which now depends upon the tides), and this would also provide an important source of domestic water in the large cities. Another important consideration for industry would be what provision would be made for locks, canals and other navigational engineering works to enable easy passage of ships through the proposed barrage.

In direct contrast, and some would argue in conflict with these urban and industrial forms of land use, the Severn Estuary is important for tourism, recreation and conservation of wildlife and natural environments. On the English side there are the seaside resorts of Clevedon, Weston-Super-Mare and Burnham-on-Sea all close to the position of the proposed Barrage, and on the Welsh side there are the resorts of Penarth and Barry. In addition to this, the Severn Estuary is an important breeding ground for many species of birds, especially waders and waterfowl and there are several wetland nature reserves, including the internationally renowned Wildfowl and Wetlands Trust at Slimbridge in Gloucestershire. Whereas the tourist industry may be enhanced at places such as Weston-super-Mare by barrage construction, the effects it might have on the wetland environments could be very negative indeed. As with any other scheme there would be gainers and losers. On the whole those likely to gain from the building of a barrage would include: those involved in the construction industry, industries that rely on large quantities of fresh water, the tourist industry, the electrical generation company involved in the project and certainly the local population that would be employed in the construction and maintenance of the project. Those most likely to give continuous opposition to the scheme are the environmentalists, some of the people engaged in the tourist industry who would not benefit from the changes, and those who have retired to the area and its coastal resorts in search of a quiet life, who might, in particular, suffer during the construction period.

3 Pollution and the Emergency Management of Coastlines

One of the biggest threats to marine environments is pollution, which can come from a variety of sources such as industrial plants, sewage, powers stations (which cause thermal pollution) and the oil industry. It is the last of these that has become one of the most vexing forms of pollution, because of the difficulties in managing oil spills and the long-term effects that may result. Huge amounts of oil are transported vast distances around the world in some of the biggest vessels ever built.

Some of the worst coastal oil spills in the last few decades have included the *Torrey Canyon* that spilled 119 000 tonnes of oil off the Cornish coast in 1967, the *Amoco Cadiz*, which spilled 230 000 tonnes off the coast of Brittany in 1978, and the *Braer*, which spilled 84 000 tonnes off the Shetland Islands in 1993. The *Prestige*, which sank off the Galician coast in 2002, was carrying 77 000 tonnes of oil.

CASE STUDY: THE *PRESTIGE* OIL SPILL IN GALICIA, SPAIN

The western coast of Galicia in north-west Spain has a rugged granite shoreline with cliffs, bays and rias. With its very clear waters and sandy beaches, it is both important for fisheries and for tourism. On 18 November 2002 the single-hulled oil tanker the *Prestige* split into two during rough seas off the coast of Galicia spilling its cargo of heavy fuel oil onto the shore. The ship was carrying a total of 77 000 tonnes of this heavily polluting form of petroleum. Oil was leaking at a rate of some 120 000 litres per day and soon formed a slick 50 km long and 18 km wide. Both the Spanish and Portuguese authorities had refused to let the ship, when it first got into difficulties, come into a safe harbour and as a result it sank into deep water about 800 km away from the mainland. Some predicted that this situation would stop further damage as the fuel oil would solidify in the cold of the deep waters and no longer leak; this was not to be the case and new polluting oil may continue to leak out until 2006.

The two most devastating effects of the *Prestige* oil spill were upon the wildlife and natural environment, on the one hand, and upon the livelihoods of the Galician people, on the other. Around 250 km of coastline was under threat from the oil spill and some 140 individual beaches became polluted. Over 15 000 seabirds were killed by the oil and many more times this figure were put in danger. Thousands of volunteers helped the emergency services in the clean-up operation, but the Spanish government was heavily criticised for its lack of action in the early days of the crisis and for misinformation. Worst hit were the Galician fisheries that employ about 21 000 people, mainly as small family operations with individual fishing boats and small crews, which had to be shut down overnight. The Galician shellfish gatherers are some of the most important in Europe and they also had to stop operations. On top of this, the shellfish and crustaceans act as part of the food chain for a huge variety of other creatures including birds, porpoises and whales, which have had to migrate elsewhere in search of supplies of fish, crustaceans and shellfish. The biggest medium-term problem is knowing just how much more oil is likely to escape from a ship that is now on the ocean floor.

One of the longer term solutions to such oil tanker accidents is much stricter control over the tankers themselves. The *Prestige* was sailing under a flag of convenience, that of the Bahamas but owned by a company based in Liberia. At the same time it was managed by a Greek company, chartered by a Russian firm and carrying the oil belonging to a British company from Latvia to Singapore, under a Greek captain but served by a Philippino crew. Under these sorts of circumstances, inspections for the need for repairs and the awarding of fines for pollution become very complex issues and no one takes the blame. Clean-up costs will eventually go into hundreds of millions of euros, yet the *Prestige* was only insured for €25 million in clean-up costs in the

case of being wrecked. One of the eventual factors that will miti-
gate against future spills of this nature will be the banning of
single-hulled oil tankers in European waters in favour of the
much safer double-hulled tankers after the year 2015. This law
was enacted following the wreck of the *Erica*, another ageing
single-hulled tanker off the shores of Brittany in 1999.

Summary Diagram

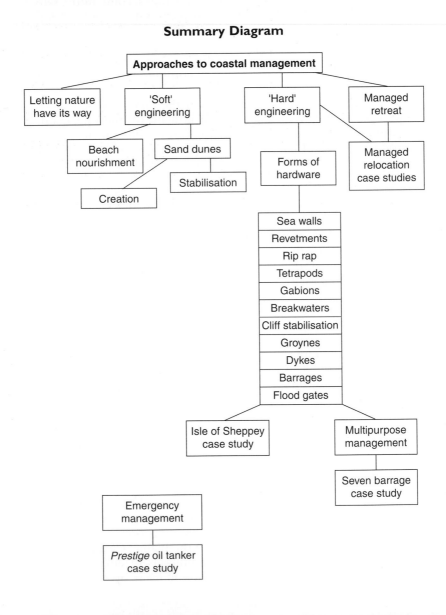

Questions

1. **a)** Outline the different approaches and attitudes that exist towards coastal management.
   b) With reference to specific places, examine the success of these different approaches.
2. **a)** Explain the differences between 'hard'- and 'soft'-engineering schemes used for coastal protection.
   b) Examine the range of 'hard'-engineering methods and their advantages and disadvantages.
3. **a)** Why are some approaches to coastal management more suitable for some locations than others?
   b) Explain why some locations are more likely to be protected than others.
4. **a)** With reference to specific examples, explain what is meant by 'multipurpose coastal management'.
   b) Describe and explain what is meant by 'managed retreat' along coastlines, considering the factors in favour and against such action.
5. **a)** Why is it increasingly important to have 'integrated coastal management'?
   b) Under what circumstances might emergency coastal management be needed?

7 Contrasting Coastlines

The river is within us, the sea is all about us;
The sea is the land's edge also, the granite
Into which it reaches, the beaches where it tosses
Its hints of earlier creation:
The starfish, the horseshoe crab, the whale's backbone;

The Dry Salvages (Four Quartets), T.S. Eliot

This chapter sets out to deal with a variety of contrasting coastlines to provide extra case study materials over and above those in previous chapters.

1 The Coastline of Somerset

The coastline of Somerset (see Figure 18) is both rich and varied. Stretching from the docks at Portbury on the edge of the built-up area of Bristol the coast trends southwards to the estuary of the River Parrett in Bridgwater Bay and then trends westwards towards the Devon border where Exmoor drops steeply into the sea and forms the highest cliffs in England. As the coastline is along the Bristol Channel it is subject to very high tidal ranges (up to 13 m during spring tides) and this has had a considerable impact on its geomorphology. A wide variety of geology is to be found along this coastline.

In the far north, between industrial Portishead and the Victorian seaside resort of Clevedon, there are cliffs that rise up to around 55 m and shore platforms both of which consist of a variety of sedimentary rocks including Devonian (Old Red) sandstone that forms almost vertical cliffs and the less-resistant conglomerates of the Triassic period that form a much more gently sloping rocky shoreline. At Battery Point in Portishead the Carboniferous limestone is also exposed.

The rest of the northern coast of Somerset is dominated by the Carboniferous limestone of the Mendip Hills. This forms a series of headlands between which there are low-lying beaches and bays. The three headlands formed from Carboniferous limestone – Sand Point, Worlebury and Brean Down, are all formed from asymmetrical anticlines and are characterised by both high and low cliffs (on Sand Point the cliffs are much higher and steeper on the northern side, and on Brean Down – they are higher and steeper on the southern side, reflecting their asymmetry), shore platforms and accumulations of pebbles and sand. Sand Point, which rises to just 48 m, and Brean Down, with its highest point at 97 m, are both owned by the National Trust and therefore in a relatively natural state. By contrast, about

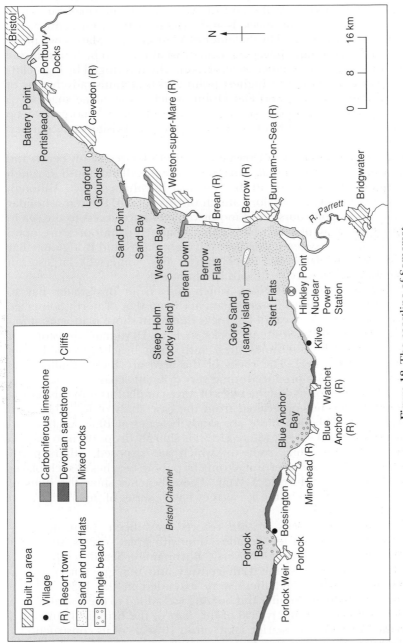

Figure 18 The coastline of Somerset

half of the Worlebury headland is part of the built-up area of Weston-super-Mare, Somerset's largest seaside resort, although the northern half is still wooded. At the end of Worlebury is the small limestone outlier that forms Birnbeck Island, which in the 19th century was joined to the mainland by the older of Weston-super-Mare's two pleasure piers. Two other limestone outlier islands are to be found in the middle of Bristol Channel: Steep Holm, which is ringed by high cliffs and reaches 78 m in its highest point and is administratively part of North Somerset UA, and Flat Holm, which, as its name suggests, is much lower in its topography and is part of South Glamorgan. Steep Holm is uninhabited but is a wildlife reserve and visitors may get permits to stay there.

To the south of Brean Down is Somerset's longest sandy beach and sand dune complex that runs from the headland south through Brean itself and Berrow to the resort of Burnham-on-Sea. Although building density is low along much of this stretch, there is a holiday camp and a golf course, and motor vehicles have access to the sea in some places, all of which have either changed or put pressure upon the natural ecosystems. The lowest lying area around Bridgwater Bay and the mouth of the River Parrett has Somerset's largest salt marshes, which were dealt with in Chapter 4.

The west Somerset coast, which runs from Bridgwater Bay to Exmoor, has a very different character from that of north Somerset. It is dominated by areas of flat rocks that form shore platforms at low tide, high cliffs of resistant rocks such as Devonian sandstone or Triassic and Jurassic limestones, and storm beaches formed from pebbles and gravel. The rocks of the area are highly faulted and folded, which has produced some very dramatic coastal scenery. The variety of cliff forms and the types of wave-cut platforms in this part of Somerset can be appreciated from the case study of Kilve beach in Chapter 3. Although there are sandy beaches at Blue Anchor and Minehead, further west where Exmoor sharply drops off into the sea there are high cliffs, some of which are exposed, others partly wooded, which are interrupted only by pebble beaches at such places as Bossington and Porlock Weir. These beaches show the signs of frequent pounding by the sea as they have a series of steep ridges and berms.

Two of Somerset's seaside resorts have been given new sea-defences in the last few decades, one with great success, the other being surrounded by controversy. Burnham-on-Sea developed as a resort at the end of the 19th century and was given its first sea-defences in 1911, and these were extended in the 1930s when a marine lake for swimming and boating was added to the seafront. The high rate of silting up, typical of this part of the Bristol Channel, led to the demise of the marine lake within 10 years of its construction.

The original sea-walls, which were of the vertical sort, were partially destroyed in a great storm surge in December 1981. New and much more hydraulically designed defences replaced the old ones and were finished by the summer of 1988. These are based upon curved walls, which can take a much greater force from storm waves and dissipate their energy; as well as walls there are revetments and stepped walls in certain places, and their design is closely related to the shape of the local topography. The sea-wall is 2 km long and, at a cost of £7.5 million, continues to do its work well.

In 1998 a press release from a construction firm was praising its own efforts for having completed a sea-defence in Minehead, well ahead of schedule. The scheme was to replace and strengthen the old sea-walls of Minehead, which had been badly damaged by storm action in 1995. Costing some £13 million, a new 1.8 km-long curved sea-wall was built and it was fronted by 100 000 tonnes of rip rap brought in by train. The new scheme, however, had the effect of removing what was left of the golden sands of Minehead's beach and left it just a mixture of mud and rocks. This caused local uproar, particularly as the tourist industry is the largest employer in west Somerset and people feared that visitors would be deterred from coming to Minehead. The construction firm have since been reassuring by shipping in 300 000 tonnes of beach material, thereby carrying out a beach replenishment strategy as stage 2 of their improvements to the seafront area.

2 The Venetian Lagoon

The Venetian Lagoon (see Figure 19), which is located on the Adriatic coast of north-east Italy, extends over some 55 000 hectares, making it one of the largest wetland areas in Europe. The seaward side of the lagoon stretches some 50 km from the popular seaside resort of Lido di Jésolo and the mouth of the River Piave in the north to the fishing town of Chióggia and the mouth of the River Brenta in the south. The lagoon was created by a mixture of coastal deposition and fluvial deposition from the rivers Piave and Tagliamento. Currents in the Adriatic and the dominant north-easterly winds have given rise to a north–south pattern of longshore drift that led to the formation of the three long, narrow, sandy barrier islands of Cavallino, Lido and Pellestrina that almost seal the lagoon off from the Adriatic. Dredged navigational channels with breakwaters at their mouths (*porti*) keep the lagoon open to the sea. The Venetian Lagoon is a complex area with many different environments and problems.

a) The different environments of the lagoon

The Venetian Lagoon is dominated by two main types of natural environments, known locally as the *laguna viva* (living lagoon) and

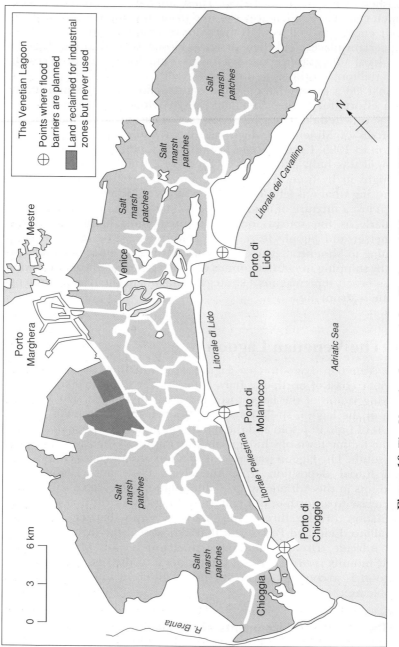

Figure 19 The Venetian Lagoon (after Tooley and Jelgersma, 1992)

laguna morta (dead lagoon). The living lagoon is predominant and is the area of shallow water where traditionally the fisheries were most active. The dead lagoon is the area made up of hundreds of intricate muddy islands and sand bars called *barene* with creeks and reed beds between them. On some of these islands there are well-established settlements, including the magnificent historic city of Venice itself, the fishing island of Burano, Murano with its glass and other craft industries, and Torcello, the former main city of the lagoon. Although the reed beds and sand banks are extremely rich in fauna, particularly bird life, very little of the lagoon is has been made into wildlife reserves (there is one small WWF run reserve at Valle Averto), most of it is now protected from further development.

b) Human activity in the lagoon

Venice was founded in the 5th century AD and rose to become a dominant European power between the 9th and the 16th centuries. The city was protected by its position in the lagoon and also became a wealthy maritime trading state because of its location. From early times the builders of Venice learnt to cope with the marshy islands by constructing the houses and churches of the city upon huge wooden piles driven into the mud and clay. Typically, the foundations of the buildings consisted of concentric circles of wooden stakes 2–4 m long, and made from the trunks of alder and larch trees being driven into the mud and clay and then thick layers of marble were 'floated' upon them giving a relatively firm base for construction. Although the buildings were deliberately kept light by having wooden frames within them, subsidence into the lagoon has been historically one of Venice's major problems. Various measures were taken in the past to try to make the situation better; for example, over several hundred years the main rivers that dumped sediments into the lagoon were diverted and this had a positive effect on the water levels. In the 18th century major sea-walls were built to protect the city, but they frequently fell into a bad state of repair.

Throughout the 20th century Venice's problems became more complex and more acute. During that century the relative sea level rose by about 33 cm – a higher rate than recorded for any previous century. Tidal and storm surges have become more frequent and the number of times there is '*acqua alta*' (high water) when the main Piazza San Marco is flooded and people have to cross it on boardwalks is also increasing. In 1900 the high water levels were reached only seven times, in 1989 they occurred 40 times and in 1996 they occurred 99 times. November 1966 saw the biggest flood for more than a century when the city was inundated by 2 m of water.

Four factors have been making the situation in the lagoon worse:

- the development of Mestre and Porto Marghera as large industrial cities on the mainland; not only did it become a source of

Acqua alta in Piazza San Marco, Venice

pollution in the lagoon but land reclamation associated with them began to upset the ecological balance within the lagoon
- further areas of reclaimed land earmarked for industrial development also affected the traditional environments of the lagoon, but luckily the industrial development did not take place
- deep water channels were dredged through the lagoon for large-scale shipping: oil tankers need to pass through to the industrial cities and the channels also give access to Venice for cruise ships; these dredged channels have upset the patterns of currents in the lagoon
- at the entrances to the deep water channels, at the Porti di Lido, di Malomocco and di Chioggia, long jetties were constructed to help to protect the channels from sedimentation and these have had side effects of upsetting the coastal currents

c) Dealing with the problems of flooding and subsidence

Whereas subsidence has to be dealt with at the level of individual buildings and building complexes, the overall problem of flooding from the Adriatic has a more far-reaching solution. Floodgates sealing off the three entrances to the lagoon at the *porti* have long been put forward as the best way of securing Venice and the other inhabited islands from sea surges in the Adriatic. Floodgate proposals have been considered since the mid-20th century, but it has only been since the increased frequency of flooding in the 1980s that there has been some sense of urgency. As with other large-scale engineering schemes in Italy, the decision-making process has been held up because of political rivalries, vested commercial interests, the problems of funding and bureaucracy. The success of the Thames Barrier in London has, to some extent, made the Venetians keener to resolve the problem.

There have been numerous different design proposals put forward over the years including various types of flap-gate, drum-shaped rotating gates as in the Thames Barrier, floating caissons and inflatable cylindrical rubber gates. The design that has finally been adopted as the best one for Venice is that of flap-gates that would normally be folded away on the bottom of the lagoon and then raised to block off the entrances to the lagoon when there is a threat of flooding. They will be made in prefabricated sections and put together at the three inlets: Chioggia will require 18 sections, Malamocco 20, and the much wider entrance at the Lido will have two gates of 20 and 21 sections that will meet in the middle. Full-scale tests have been carried out with flap-gate sections and they have proved to be a success. The project, once started, will take up to 8 years to complete, create over 1000 jobs and is likely to cost well over €2 billion.

Individual buildings in Venice are now being restored and stabilised from further subsidence by a wide variety of organisations. The sheer number of palaces, churches and other buildings of world importance have made it impossible for the local authorities to carry out these changes on their own. Since the great flood of 1966, Venice has become the responsibility of the whole world. In 1966 UNESCO (the United Nations Educational, Scientific and Cultural Organisation) launched an appeal to save Venice and over 50 private organisations such as the British 'Venice in Peril' fund came forward and adopted individual buildings for restoration and other projects to save the city. Eleven different MEDCs are involved in these projects and work will continue for many more decades.

3 Singapore Island: Reclamation and Stabilisation

Singapore is an island republic that has undergone a vast transformation in the last 40 years. Its biggest problems have in this period been connected with restriction upon space. The country has a population of just over 3.5 million but has a land area of just 646 km², making it one of the most densely populated nations on earth, with almost 5500 people per km². With the rapid population growth and the economic boom over the last few decades Singapore has been pressurised by the need for new land for housing, and for both heavy and light industries, as well as infrastructure. Until the early 1970s most of the new land came from the appropriation of farmland, but since the 1970s it has come through land reclamation. Extensive reclamation projects have been carried out around both Singapore island itself and around the smaller offshore islands (see Figure 20). Given present technology the reclamation projects are limited to seas with a depth of less than 15 m. When all the current plans have come to fruition Singapore will have a total area of 736 km², which will be an increase of roughly 25% on its land area in 1967.

Reclamation is expensive to carry out as it involves building dams around the sites being reclaimed and massive amounts of infill. In the past it was domestic and other forms of raw waste used for this infill, now most of the material is incinerated beforehand and the ash is used along with natural materials dredged from the sea.

Land reclamation is concentrated in certain areas around Singapore, as can be see in Figure 20.

(i) In the city centre, where most of the new high-rise commercial buildings are now on reclaimed land. One of the most recent new projects to open in this area is the Esplanade concert hall and theatre complex. Future plans will see a huge extension of the reclaimed land in the city centre, particularly in the Marina Bay area.

(ii) Around Changi Airport. The airport itself and its recent extension are on reclaimed land. At present a huge new area of reclamation is taking place near the airport to provide new port facilities taking some of the pressure off the main port of Singapore.

(iii) At Jurong and Tuas. This is Singapore's main industrial zone with both heavy processing plants and high-tech industries located on reclaimed land. The main oil refineries and associated chemical works are located on offshore islands near Jurong; future plans would consolidate several smaller offshore islands into one large Jurong Island.

(iv) In resort areas such as the East Coast Park, sand has been brought in to create a series of leisure beaches for the Singaporean population. Sentosa Island, Singapore's main resort area, has also been extended. There are new beaches being created on reclaimed land around Kusu and Lazarus Islands in the south.

At the same time that all the reclamation is taking place, Singapore has to cope with the destruction of beaches. Some of this is taking place along the 8-km stretch of the East Coast Park and some in the Pasir Ris resort area on the north coast; both of these are artificial resorts built since the 1970s using imported beach materials from elsewhere. Wave energy is low in Singapore – the average wave height is just 0.6 m; at the same time, fetches are short because the main island is protected by the Malay Peninsula immediately to the north and by small groups of islands, including some of the reclaimed and industrialised ones to the south. The main problems facing the

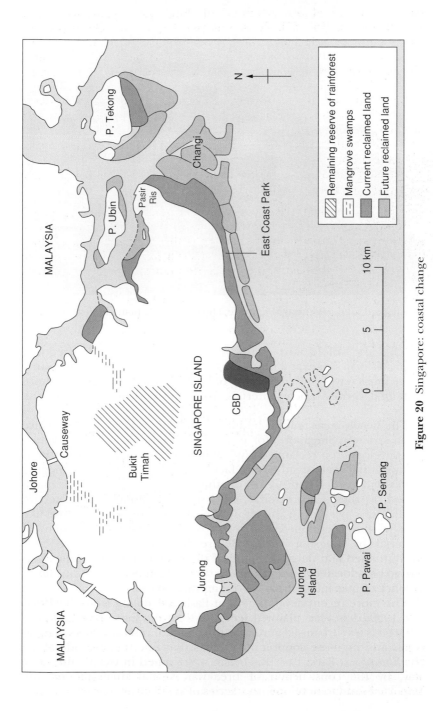

Figure 20 Singapore: coastal change

Reclaimed land in central Singapore, showing the Esplanade development

beaches are twofold: wave energy is much higher during the frequent tropical storm surges and this leads to beach erosion, and, at the same time, most waves are coming in obliquely, which causes longshore drift that carries beach materials from one end of a beach to the other.

The beach areas tend to be highly built up as a result of pressure upon space in Singapore, for example the East Coast Resort, which was developed on reclaimed land. When the resort was first set up it had over 100 chalets and leisure facilities such as cycle tracks, jogging tracks and barbecue pits close to the sea. The beaches were maintained, as they still are today, by a series of stone-built protective breakwaters parallel to the shoreline. Against these small cusp-like beaches have developed as a result of the east–west movement of material by longshore drift. Before the breakwaters were later strengthened and enlarged some of the chalets and leisure facilities had to be relocated as a result of the threat from beach erosion. Thus, a policy of managed retreat was being carried out.

The more recent development at Pasir Ris, dating from the 1980s and 1990s, has a less planned feel to it than the East Coast Park. For a start it is less urbanised and lacks the chain of restaurants, swimming pools and high-rise condominium flats that line the roadside along the East Coast Park. Pasir Ris beach was formed in exactly the same way, by the construction of breakwaters that then allowed the imported sand to develop into a series of small cusped beaches by the natural process of longshore drift. Various holiday flats and chalets are located close to the beach but they give the area a much more rural feel than along the East Coast Park. Patches of mangrove swamps have been preserved along this stretch of coastline and along the Tampines River, which drains out into the sea at Pasir Ris; board-walks and a nature trail have been created through the mangroves.

The removal of the extensive mangrove swamps, which once lined the island's coasts, is one of the main reasons why the beaches are so vulnerable to wave attack. Mangrove swamps harboured malarial mos-quitoes and it was therefore in the interests of the British colonial rulers and then, after independence in the early 1960s, for the Singaporean authorities to clear the swamps for reasons of public health. Although malaria has been eradicated in Singapore, dengue fever, also spread by mosquitoes, is still a problem and a reason why mangrove swamps can still be a public health risk. The other unsavoury aspect of mangrove swamps was that they were an easy place to dump domestic and other rubbish – also a threat to public health in a hot, humid tropical climate.

Mangroves were widely cleared in the various coastal reclamation schemes, so that now there are only a few small patches along the coast where they survive. One of the most extensive remaining areas of mangrove swamp in Singapore is the 87-hectare Sungei Buloh Nature Park in the north of the main island. This provides boardwalks and nature trails for paying visitors and has an important educational function.

The mangrove swamps with their dense vegetation are able to absorb a great deal of wave energy and thereby hold beach materials together and to protect the coastline from erosion, even during storm surges. With the disappearance of the mangroves, the Singaporean authorities have had to invest millions of dollars in the building of hard structures such as sea-walls and breakwaters in order to protect the shoreline from erosion and longshore drift. Since the mid-1990s there has been a bit of a change in the attitude of the authorities, with engineers and planners re-assessing the role of mangroves along the coasts of Singapore. Now people are looking favourably towards mangroves, and certain areas have been set aside for mangrove conservation and regeneration.

Summary Diagram

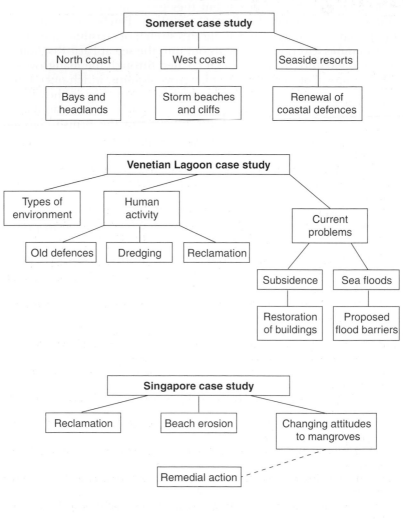

Questions

1. **a)** With reference to selected places that you have studied, explain why there can be great variations in landforms and scenery along relatively short stretches of coast.

    b) How do these different types of coastal landscapes affect human activity?

    (As this chapter has additional case study material, it should be used to help answer the questions at the end of the other chapters.)

Bibliography

Bird, E., 2000, *Coastal Geomorphology: An Introduction* (Chichester: Wiley)

Briggs, D. *et al.*, 1997, *Fundamentals of the Physical Environment* (London: Routledge)

Bristow, C., 1996, *Cornwall's Geology and Scenery* (St Austell: Cornish Hillside Publications)

Castledon, R., *Classic Landforms of the Sussex Coast* (Sheffield: Geographical Association)

Clayton, K., 1979, *Coastal Geomorphology* (London: Macmillan)

Crane, W., 1992, *The Environment of Tonga* (Lower Hutt, NZ: Wendy Crane Books)

Davis, R., 1996, *Coasts* (Upper Saddle River, New Jersey: Prentice Hall)

Fam, V., 2001, *Interactive Geography 3* (Singapore: Pan Pacific Publications)

French, P., 1997, *Coastal and Estuarine Management* (London: Routledge)

Goudie, A., 1984, *The Nature of the Environment* (Oxford: Blackwell)

Goudie, A., 1990, *The Landforms of England and Wales* (Oxford: Blackwell)

Goudie, A. and Brunsden, D., 1994, *The Environment of the British Isles: An Atlas* (Oxford: OUP)

Goudie, A. and Gardner, R., 1992, *Discovering Landscape in England and Wales* (London: Chapman & Hall)

Hallam, J. *et al.*, 1984, *Earthshaping 4: Coastmaking* (Slough: UTP)

Hardy, P., 1999, *The Geology of Somerset* (Bradford on Avon: Ex Libris Press)

Haslett, S., 2000, *Coastal Systems* (London: Routledge)

Hill, M., 1999, *Advanced Geography Case Studies* (London: Hodder & Stoughton)

Morton, B. *et al.*, 1998, *Coastal Ecology of the Açores* (Ponta Delgada: Sociedade Afonso Chaves)

Mottershead, D., 1986, *Classic Landforms of the South Devon Coast* (Sheffield: Geographical Association)

Pethick, J., 1984, *An Introduction to Coastal Geomorphology* (London: Edward Arnold)

Pinna, M. and Ruocco, D. (Eds), 1980, *Italy: A Geographical Survey* (Pisa: Pacini)

Polunin, O. and Walters, M., 1985, *A Guide to the Vegetation of Britain and Europe* (Oxford: OUP)

Scarth, A. and Tanguy, J.-C., 2001, *Volcanoes of Europe* (Harpenden: Terra Publishing)

Small, R., 1989, *Geomorphology and Hydrology* (London: Longman)

Steers, J., 1969, *Coasts and Beaches* (Edinburgh: Oliver & Boyd)

Tooley, M. and Jelgersma, S. (Eds), 1992, *Impacts of Sea-level Rise on European Coastal Lowlands* (Oxford: Blackwell)

Trueman, A., 1971, *Geology and Scenery in England and Wales* (Harmondsworth: Penguin)

Williams, M. (Ed.), 1990, *Wetlands: A Threatened Landscape* (Oxford: Blackwell)

Websites

www.defra.gov.uk; www.environment-agency.gov.uk – these two are the most important organisations concerned with coastal flooding in Britain.

www.geographyinthenews.rgs.org – this website will have any important coastal issues in it as and when they happen.

www.igu-net.org – this has clickable links to any university geography department in the world.

www.usgs.gov – this is the main organisation responsible for action in monitoring coastal changes in the USA.

www.soest.hawaii.edu; www.coastal.tamug.edu – these are two excellent websites dealing with coastal changes in individual US states, the first for Hawaii the second for Texas. All US states with coasts have similar sites for educational use.

Index